Springer-Verlag Berlin Heidelberg GmbH

Rüdiger Schade Irene Behn Michael Erhard
Andreas Hlinak Christian Staak (Eds.)

# Chicken Egg Yolk Antibodies, Production and Application

## IgY-Technology

With 62 Figures, 3 in Color

Springer

PD Dr. RÜDIGER SCHADE
Universitätsklinikum Charité
der Humboldt Universität zu Berlin
Institut für Pharmakologie
und Toxikologie
Dorotheenstraße 94
10117 Berlin, Germany

Dr. INGRID BEHN
Universität Leipzig
Fakultät für Biowissenschaften,
Pharmazie und Psychologie
Institut für Zoologie
Talstraße 33
04103 Leipzig, Germany

Prof. Dr. MICHAEL ERHARD
Universität Leipzig
Veterinärmedizinische Fakultät
Veterinär-Physiologisches Institut
An den Tierkliniken 7
04103 Leipzig, Germany

Dr. ANDREAS HLINAK
Freie Universität Berlin
Veterinärmedizinische Fakultät
Institut für Virologie
Luisenstraße 56
10117 Berlin, Germany

Prof. Dr. CHRISTIAN STAAK
Bundesinstitut für gesundheitlichen
Verbraucherschutz und Veterinär-
medizin (BgVV)
Institut für Veterinärmedizin, FG 505
Diedersdorfer Weg 1
12277 Berlin, Germany

ISBN 978-3-662-04490-2 ISBN 978-3-662-04488-9 (eBook)
DOI 10.1007/978-3-662-04488-9

Library of Congress Cataloging-in-Publication Data
Chicken egg yolk antibodies, production and application: IgY-Technology/Rüdiger Schade ... [et al.].
p. cm. - (Springer lab manuals) Include bibliographical references and index. ISBN 978-3-662-04490-2 (alk. paper) 1. Immunoglobulins - Laboratory manuals. 2. Eggs - Laboratory manuals. 3. Chickens as laboratory animals - Laboratory manuals. I. Schade, Rüdiger. II. Series.
QR186.7.C48 2000 616.07'98 - dc21



Production: PRO EDIT GmbH, 69126 Heidelberg, Germany
Cover design: design & production GmbH, 69121 Heidelberg, Germany

# Preface

As early as 1893 Klemperer published his observation that there must be neutralising proteins (i.e. antibodies) in the yolk of eggs laid by immunised hens. As is often the case in the history of science, this finding was hardly acknowledged and even ignored for a long time. Only later, when animal protection came to be regarded as a serious ethical claim for the scientific community as well, did these results came back to mind. Animals are now perceived as fellow creatures in need of protection, so the extraction of specific antibodies from the eggs of immunised hens was regarded as an attractive alternative to blood-taking methods.

In 1992 – nearly a century after the original discovery – yolk immunoglobulins became the topic of a joint research project. Various research groups which already had experience with yolk antibodies started to work together and were financially supported by the German Ministry for Education and Research. The aim of this project was to find out whether these avian antibodies are as effective as the traditionally used polyclonal antibodies from mammals, especially from rabbits. It could be shown that in the case of phylogenetically highly conserved antigens, avian antibodies are even more effective than mammalian antibodies.

The project went on for six years, during which IgY-technology won increasing attention and acceptance. There are now commercially available cages for keeping hens under acceptable conditions. In addition to laboratory protocols for extracting IgY from yolk, there are also easy-to-use commercial kits for the isolation of immunoglobulins. Also, secondary antibodies coupled to detectable markers are now available. Within a few years, IgY-technology has developed to a recognized alternative which is less stressful to animals than the conventional procedures. In other words, there is no longer any excuse for not using this well established technology.

In acknowledgement of their work on yolk antibodies, the scientists working together in the joint project received the FISEA prize in1997. FISEA – the International Foundation for the Substitution of Animal Experimentation, Luxemburg – awards this prize to researchers working on new alternative methods that eleminate or reduce the use of animals or are less stressful to animals.

Apart from the animal protection aspect, there are some advantages in using yolk antibodies compared to mammalian. For instance, a single immunised hen can yield an amazing abundance of antibodies, many of which show a different specificity than those a rabbit would produce after the same immunisation. These benefits have already led to scientific results which could not be achieved with mammalian antibodies, neither with poly- nor monoclonals.

The only obstacle was the lack of a reliable collection of protocols and laboratory instructions for this technology. Researchers or students who wished to apply this technique, were still obliged to search for information from literature in various fields of specialisation. This book now presents all relevant protocols and information needed to use IgY-technology easily and successfully.

The protocols are based on our own experience and on those of our colleagues working in the same field. All procedures have been tested and most have already reached a routine level. They are presented in an easy-to-follow way, supported by many practical hints and notes. Included are detailed instructions on how to keep and immunise hens and protocols for the extraction of antibodies from yolk and for the various applications. The protocols were selected with the aim to cover as broad a spectrum as possible. In some cases, it will of course, still be necessary to refer to special literature. Therefore, a list of the relevant references is given in the appendix.

We do hope that this laboratory manual will enable also the unexperienced scientist and everybody working in a laboratory to use this technology without initial problems.

We are aware that one can never avoid some mistakes finding their way into a text and wish to apologise for these possible errors in advance. We would welcome any suggestions which support the book's aim to disseminate IgY-technology and thus serve the protection of animals.

Berlin, Leipzig
Juli 2000

PD Dr. RÜDIGER SCHADE
Dr. INGRID BEHN
Prof. Dr. MICHAEL ERHARD
Dr. ANDREAS HLINAK
Prof. Dr. CHRISTIAN STAAK

# Acknowledgement

It is a great pleasure for the editors to express their sincere gratitude to the subsequently named official institutions, sponsors, private institutions, foundations for their motivating and/or financial support. This support made it possible to edit this laboratory manual on IgY-technology.

- Bundesministerium für Bildung und Forschung, Bonn, Germany
- Projektträger BEO, Jülich, Germany
- Zentralstelle zur Erfassung und Bewertung von Ersatz- und Ergänzungsmethoden zu Tierversuchen (ZEBET) im Bundesinstitut für gesundheitlichen Verbraucherschutz und Veterinär-Medizin, Berlin, Germany
- Stiftung zur Förderung der Erfassung von Ersatz- und Ergänzungsmethoden zur Einschränkung von Tierversuchen, Mainz, Germany
- Erna-Graff-Stiftung für Tierschutz, Berlin, Germany
- FISEA, Luxembourg
- Universitätsklinikum Charité, Berlin, Germany
- Vet.-Medizinische Fakultät der FU (Standort Mitte), Berlin, Germany
- EBECO, Castrop-Rauxel, Germany
- BioGenes GmbH, Berlin, Germany
- ehemalige Landwirtschaftliche Lehr- und Versuchsstation der Humboldt-Universität zu Berlin – SPF Hühnerhaltung (Leiter Dr. H. Kobilke), Damsdorf, Germany

The editors are greatly indebted to Mrs. H. Glatte for the preparation of figures and to Mr. P. Stanway for the translation, and thus for their valuable assistence in preparing the manuscript.

# Abbreviations

| | |
|---|---|
| Ab | antibody |
| ABTS | 2,2'-azino-bis(etylbenzthiazoline-6 – sulfonic acid) |
| AGIDT | agargel double immunodiffusion test ("Ouchterlony"-technique) |
| AI | avidity index |
| ALP | alkaline phosphatase |
| BGG | bovine gamma globuline |
| BSA | bovine serum albumine |
| CCK | cholecystokinin |
| CD | cell differenciation |
| CRP | C-reactive protein |
| DAB | diaminobenzidine |
| DMSO | dimethylsulfoxid |
| DNP | 2,4-dinitrophenol |
| ECVAM | European Centre for the Validation of Alternative Methods |
| EIA | enzyme immunoassay |
| ELISA | enzyme-linked immunosorbent assay |
| FACS | fluorescence activated cell sorter |
| FCA | Freund's complete adjuvans |
| FIA | Freund's incomplete adjuvans |
| FITC | fluorescein isothiocynate |
| GABA | $\gamma$-aminobutyric acid |
| HPR | horseradish peroxidase |
| HSA | human serum albumin |
| i.d. | intradermal |
| IE | immunoelectrophoresis |
| IEA | immunoelectrophoretic assay |
| IgA | immunoglobulin A |
| IgG | immunoglobulin G |
| IgM | immunoglobulin M |

| | |
|---|---|
| IgY | immunoglobulin Y |
| i.m. | intramuscular |
| i.p. | intraperitoneal |
| ISCOM | immune stimulating complex |
| i. v. | intravenous |
| kDa | kilo Dalton |
| KLH | keyhole limpet hemocyanine |
| LPS | lipopolysaccharide |
| mAb | monoclonal antibody |
| MAP | multiple antigen peptide |
| MBP | myelin basic protein |
| MDP | muramyldipeptide |
| MHC | major histocompatibility complex |
| OPD | ortho-phenylenediamine |
| PAGE | polyacrylamide gelelectrophoresis |
| PBS | phosphate-buffered saline |
| PCSL | $Pam_3Cys$-Ser-$(Lys)_4$ – lipopeptide |
| PEG | polyethylenglycol |
| pNPP | p-nitrophenylphosphate |
| POD | peroxidase |
| RIA | radio immunoassay |
| RIE | "rocket" immunoelectrophoresis |
| rpm | revolutions per minute |
| RZ | Reinheitszahl |
| s.c. | subcutaneous |
| SD | standard deviation |
| SDS | sodium dodecylsulfate |
| SoAg | somatic antigen |
| SPF | specific pathogen free |
| SRID | single radial immunodiffusion ("Mancini"-technique) |
| TBS | tris-buffered saline |
| TMB | tetramethylbenzidin |
| Tris | Tris-hydroxymethyl-aminomethan |
| TRITC | tetramethylrhodamine isothiocynate |
| v/v | volume/(total)-volume |
| w/v | weight/ (total)-volume |

# Contents

# Short Introduction to Hens' Humoral Immune System

MICHAEL ERHARD and RÜDIGER SCHADE

## Introduction

The hens' immune system differs from mammals' in various ways. In this brief introduction we shall try to give a basic idea of the special features of the hens' humoral immune system, especially as regards the structures and functions of antibodies.

Like the immune system of mammals, that of hens is divided into two major components, one of them being non-specific and innate and the other specific and acquired. Naturally between the two there are numerous interactions of critical importance in the event of an immune response. Hens too, for instance, have antigen-presenting cells, and many functional mechanisms are regulated via interleukines. Unfortunately there are few data about hens, since there has been less research internationally into the immune system of hens than into that of rats and mice. But hens do have a feature not shared by mammals, as Klemperer (1893) pointed out: the passive immunity of the offspring, which in mammals reaches the foetus through the placenta or colostrum, has to come through the fluid parts of the egg in ovipars. While the egg is still in the ovary, hens transfer their serum immunoglobulin Y (IgY) into the yolk. As the egg passes down the oviduct, IgM and IgA from oviduct secretion are required with the albumin.

---

✉ Michael Erhard, Universität Leipzig, Veterinärmedizinische Fakultät, Veterinär-Physiologisches Institut, An den Tierkliniken 7, Leipzig, 04103, Germany (*phone* +49-341-97-38062; *fax* +49-341-97-38096; *e-mail* erhard@rz.uni-leipzig.de)
Rüdiger Schade, Universitätsklinikum Charité der Humboldt-Universität, Institut für Pharmakologie und Toxikologie, Dorotheenstr. 94, Berlin, 10117, Germany

So hens' eggs as a source of antibodies (Lösch et al. 1986) invite a wide variety of questions. We shall be brooding over them in chapters 6 - 8 of this handbook on methodology.

## General basis of hen's humoral immune response

Immune competent avian cells of the specific immune system come from the thymus (T-lymphocytes) or from the bursa of Fabricius (B-lymphocytes). In the course of embryonic development, precursors of these cells reach the thymus or the bursa and there differentiate into T- and B-lymphocytes, as in mammals.

The humoral immune response, in the form of antibody-synthesis, is an important part of the specific response to foreign structures (antigens). According to the clone selection theory from Burnet (1959), an activated B-lymphocyte (plasma-cell) is genetically designed to produce an exactly defined antibody, which is firstly expressed as an antigen receptor (B-cell receptor - BCR) on its cell membrane. The BCR can bind to unprocessed antigen molecules. After proliferation induced by antigens, these B-lymphocytes differentiate into so-called effector cells, which are plasma cells producing antibodies. Antibodies are simply soluble forms of BCR secreted into the body fluids. Every antibody binds only one specific epitope of an antigen, also known as an antigen determinant. So antigens are molecules against whose surface-determinants a vertebrate is able to produce antibodies. This in turn presupposes that the respective antigen structure is recognised as foreign. Not every antigen is immunogen for every individual, so highly conserved mammal antigens from hens may set off a specific immune reaction without being immunogen to the mammal itself.

In general, a distinction has to be drawn between antigens and antigen determinants. All kinds of macro-molecules such as proteins and polysaccharides and hence glycoproteins, nucleoproteins, lipoproteins but also lipids, steroids or nucleic acids may act as antigens. The best antigens are proteins and polysaccharides, whereas lipids and nucleic acids are much less clearly recognised by the immune system. The antigen determinant epitop of an antigen is actually the site where the antibody binds the antigen and makes up only a small part of the whole molecule. Antigen determinants of proteins may, for instance, be made up of 3 to 10 amino acids.

On an antigen, the number of epitopes able to bond depends on the antigen's structure and size. Generally there is a difference between sequential determinants, such as one defined by a sequence of amino acids, and determinants defined by a configuration. Thus antigens can be typified immunologically by antibodies. Small molecules such as haptens, which are not themselves immunogen, can stimulate lymphocytes to synthesise antibodies, if these haptens are coupled to larger molecules serving as carriers. So in the case of hens too, antigens range from proteins, viruses, bacteria and parasites to the cells of organisms. The potency of an antigen depends in turn on the size of the molecule, and few molecules with a molecular weight below 10,000 Daltons can induce an immune response without a carrier. The structure of the antigen is always the crucial factor. Moreover, non-specific stimulating substances, known as adjuvants, can positively influence the immunogenicity.

### Characterisation of antibody-antigen binding

Antibodies are typified according to their specific responses and related affinities. The human immune system seems to have antibodies with about $10^7$ to $10^8$ specific responses. The scope of the hens' immune system is not assessed in the literature, but in hens too, each plasma cell synthesises precisely defined molecules, which are identical antibodies with their specific biological function. Raised to the level of immunological dogma, this axiom has been the prerequisite for the production of monoclonal antibodies. Every hybridoma cell clone comes from one B-lymphocyte stimulated by an antigen and accordingly produces a precisely defined antibody, the monoclonal one. Unfortunately, no suitable myeloma cell lines are presently available as fusion-partners in the hens' system. Polyclonal serum-antibodies on the other hand consist of a mixture of antibodies of various classes with various specificities and affinities, since they have come from various B- cell clones. Antigen-antibody binding is reversible in accordance with the law of mass action. The binding force between an antibody and antigen is taken as a measure of the antibody's affinity with the respective antigen.

There are no known differences between antibodies from mammals and hens in this respect.

## Immunoglobulins in hens

Antibody molecules are glycoproteins called immunoglobulins. Hens' immunoglobulins are like mammals' in having light (L) and heavy (H) chains, bridged by disulphide bonds. The molecule is made up of a variable part with an antigen binding site and a constant part. In the case of hens, a distinction is made between the immunoglobulins M (IgM), Y (IgY) and A (IgA). Hens' IgM has the same function as mammals' IgM. Present in all vertebrates, IgM delivers the first response with its high molecular weight. Likewise avian IgA is not essentially different to mammals' IgA, as it too is found in the fluids of the gall bladder and in secretions active in the body. Already in 1969, Leslie and Clem suggested that hens' main serum-immunoglobulin should be called IgY instead of IgG, and the results of more recent genetic research suggest that the IgY molecule is phylogenetically a progenitor of mammals' IgG and IgE. According to Warr et al. (1995) this conclusion is justified by the antibody's function and structure and the corresponding expression of the H-chain genes. IgY, like mammals' IgG, is the immunoglobulin delivering the second response with its high serum-concentration and low molecular weight. But in the case of hens, IgY can also take part in anaphylactic reactions unlike mammalian IgG.

Structurally there is a clear difference between IgY and mammals' IgG, as the heavy chain of hens' IgY has an additional constant domain instead of the hinge region of mammals' IgG. So the molecular weight of IgY (about 190 kDa) is rather more than IgG's (Figure 1). In the literature, the terms IgG and IgY are sometimes used as synonyms as regards hens, so on the basis of the newest findings it has been decided within the framework of an international ECVAM workshop that the term IgY should be used throughout (Schade et al. 1996).

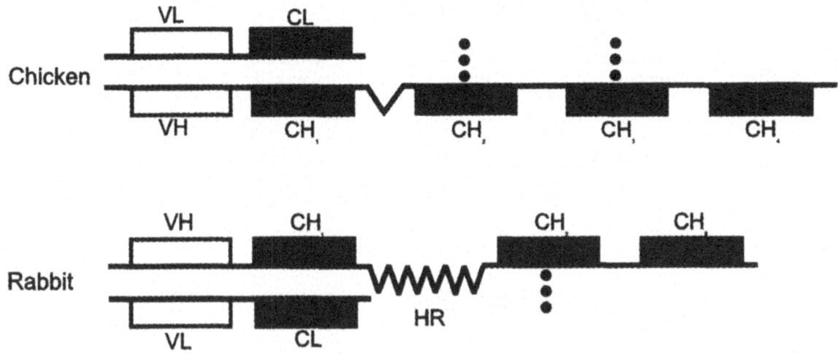


**Fig. 1.** A simplified model of a mammalian (rabbit IgG) immunoglobulin is compared with one of an avian (hen IgY) immunoglobulin. Differences are revealed in the number of the constant domains of the heavy chains (HC), the carbohydrate chains (black points) and the hinge region (HR), which is much more evident in the case of IgG. L = light chain, H = heavy chain, V = variable domain, C = constant domain. (The figure is simplified from Shimizu et al. 1992)

## Molecular basis of antibody diversity in hens as compared to those in mammals

The complete genetic arsenal of a mammal or bird would not suffice to produce so huge a variety of antibodies if every antibody molecule were coded by a separate gene. In fact only a fraction of the genes available are engaged in coding antibodies.

According to Weill and Reynaud (1996), there are three basic mechanisms behind this diversity:

- **Gene rearrangement:** site specific recombination involving recognition and cutting of heptameter-nonamer recombination signal sequences flanking the rearranging elements.
- **Gene conversion:** non-reciprocal homologous recombination whereby modifications in an acceptor gene are copied from a homologous donor sequence (templated modification).
- **Somatic mutation:** single (or double) nucleotide substitution that is not derived from a pre-existing sequence (untemplated modification).

The main device in mammals consists in rearranging various gene segments to produce the hyper-variable part of the anti-

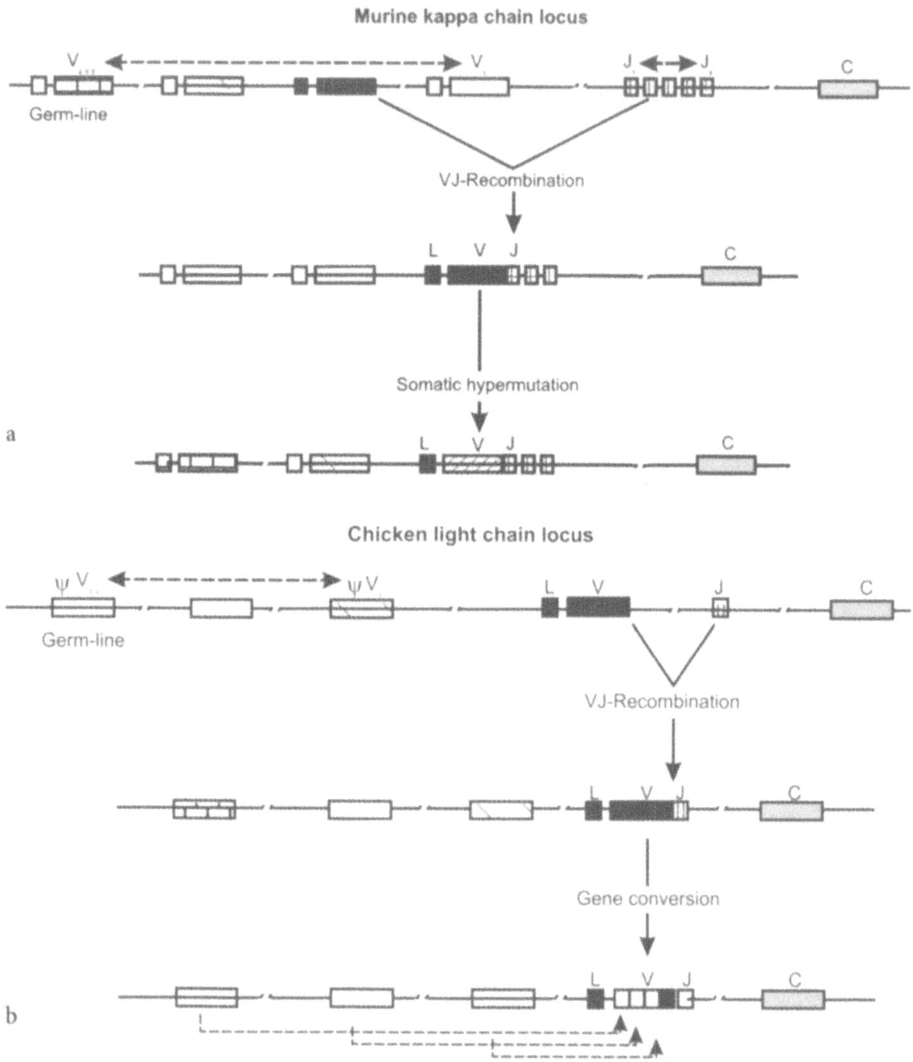


**Fig. 2.** **a** Diversification of the mouse kappa chain locus (light chain). The locus consists of about 200-1000 V-gene segments (which encode for the V-domain), 5 J-gene segments (which encode for a certain part of the V-domain) and a C-gene segment (which encodes for the constant domain). Diversification results from the range of possible VJ-combinations and additional somatic mutations. L = leader sequence. This Figure has been adapted from Bezzubova and Buerstedde (1994). **b** Diversification of the hen light chain locus. The locus consists of a single functional V, J and C gene segment. In addition there are 25 so-called pseudo V-gene segments, lacking the usual transcription regulatory and signal-recognition sequences. Diversification in the case of hens results from incorporating between 10 and more than 120 base pairs from the pseudo V-genes in the functional V-gene

body molecule and hence a spate of mutations, so antibody molecules have millions of possible variations (see Figure 2a).

Hens achieve this diversity through gene conversion. The H- and L-chain-loci are situated uniquely, as they have only a single functional $V_H$- or $V_L$- element, so on both strings of the DNA hens have a huge family of V-pseudogenes responsible for the diversity of antibody molecules (Figure 2b).

## Biological basis of immunoglobulin-transfer into eggs

Immunoglobulin in hens' eggs is transferred in various ways. IgA and IgM are transferred together with other proteins in the oviduct into the egg white. But so little is transferred that in the finished egg there are only traces (μg-magnitude), and only in the egg white. On the other hand IgY in the egg follicle is passed on by receptors, hence selectively, in large amounts into the yolk. The IgY-concentration in the yolk is comparable to the concentration of IgY in the serum (6 to 13 mg/ml). The transovarial passage takes about 5 days (Patterson et al., 1962). Both the content of IgY as a whole and the specific amounts of IgY in the serum and yolk respectively correlate positively if the time of passage into the yolk is taken into account (Erhard et al., 1997). Thus the course of titration after immunisation can be determined directly in the yolk.

The thing to remember is that the desired highly affine antibody of the second response (IgY) can be isolated in large amounts from the egg yolk without IgM- or IgA-'impurities'.

## Advantages of avian antibodies

Avian antibodies are being used more and more in basic research. The number of publications about them in recent years has risen steadily.

Due to the phylogenetic distance between birds and mammals, hens' antibodies in particular have clear advantages over mammalian immunoglobulins. For instance, highly conserved mammal-specific antigens from hens can be recognised as foreign. In part, more epitopes on mammalian antigens can be recognized from hens' antibodies than on those from antiserums from mammals. Furthermore, hens' antibodies do not bind with mammalian Fc-receptors or rheumatoid factors nor with proteins A or G, so the use of hens' antibodies is seldom likely to produce falsely positive reactions in certain immunochemical assays. Moreover a combination with mammalian antibodies is possible and sensible. Thanks to the transfer from the plasma through the egg follicle to the yolk, the amount of avian IgY available is enormous. A single hen's egg may yield about 100 to 250 mg of IgY. Thus, as this replaces blood from mammals, cruelty to animals is minimised. The only unpleasantness for hens is the immunisation, so in efforts to 'refine, reduce and replace' the exploitation of animals, the production of antibodies via hens' eggs deserves increasing attention.

## References

Bezzubova OY, Buerstedde J M (1994). Gene conversion in the chicken immunoglobulin locus: A paradigm of homologous recombination in higher eukaryotes. Experientia 50: 270-276

Burnet FM (1959) The clonal selection theory of acquired immunity. Cambridge University Press

Erhard MH, Schmidt P, Hofmann A, Bergmann J, Mittermeier P, Kaufmann P, Wiesmüller K-H, Bessler W, Lösch U (1997) The lipopeptid $Pam_3Cys$-Ser-$(Lys)_4$: an alternative adjuvant to Freund's adjuvant for immunisation of chicken to produce egg yolk antibodies. ATLA 25: 173-181

Klemperer F (1893) Ueber natürliche Immunität und ihre Verwerthung für die Immunisirungstherapie. Arch Exptl Pathol Pharmakol 31: 356-382

Leslie GA, Clem LW (1969) Phylogeny of immunoglobulin structure and function III. Immunoglobulins of the chicken. J Exp Med 130: 1337-1352

Lösch U, Schranner I, Wanke R, Jürgens L (1986) The chicken egg, an antibody source. J Vet Med B 33: 609-619

Patterson R, Younger S, Weigle WO, Dixon FJ (1962) Antibody production and transfer to egg yolk in chickens. J Immunol 89: 272-278

Schade R, Staak C, Hendriksen C, Erhard MH, Hugl H, Koch G, Larsson A, Pollmann W, Regenmortel M, Rijke E, Spielmann H, Steinbusch H, Straughan D (1996) The production of avian (egg yolk) antibodies: IgY. ATLA 24: 925-934

Shimizu M, Nagashima H, Samo K, Hashimoto K, Ozeki M, Tsuda K, Hatta H (1992) Molecular stability of chicken and rabbit immunoglobulin G. Biosci Biotech Biochem 56: 270-274

Warr W, Mayor KE, Higgins DA (1995) IgY: clues to the origin of modern antibodies. Immunol. Today 16: 392-398

Weill J-C, Reynaud, C-A (1996) Rearrangement/hypermutation/gene conversion: when, where and why? Immunol Today 17: 92-97

# Possibilities of Keeping Laying-Hens to Obtain Antibodies

ANDREAS HLINAK and RÜDIGER SCHADE

## 1 Regulations and Recommendations for Keeping Laying-Hens

Within the field of biological-medical research, laying-hens are being kept increasingly for experimental purposes, but they have a different status to most other laboratory animals in counting as farmyard-animals. There are no defined regulations as regards hens for the laboratory but there are basic regulations for the protection of animals as well as national and European regulations, guidelines and recommendations for hen-keeping for experiments:

- The Law for the Protection of Animals (Version of 25.05.1998, Germany)
- Guidelines for the Size of Cages and Coops for Keeping Domestic Hens for Experiments. Bundesamt für Veterinärwesen (National Office for Veterinary Medicine), 3097 Liebefeld-Bern, 17th Jan., 1994 (2) d. Vo/Oe/Fr/St/re-800. 116.2.04, (Switzerland)
- European Convention for the Protection of Vertebrate Animals used for Experimental and other Scientific Purposes. Council of Europe, Strasbourg 31.05.85
- The Accommodation of Laboratory Animals in Accordance with Animal Welfare Requirements (ed. P.N. O'Donoghue).

✉ Andreas Hlinak, Universitätsklinikum Charite der Humboldt-Universität, Institut für Pharmakologie und Toxikologie, Dorotheenstr. 94, Berlin, 10117, Germany (*phone* +49-30-2093-6206; *fax* +49-30-2093 4608; *e-mail* ruediger.schade@charite.de)
Rüdiger Schade, Universitätsklinikum Charite der Humboldt-Universität, Institut für Pharmakologie und Toxikologie, Dorotheenstr. 94, Berlin, 10117, Germany

Proceedings of an International Workshop held at the Bundesgesundheitsamt (National Health Office), Berlin, 17th-19th May, 1993, Bundesministerium für Ernährung, Landwirtschaft und Forsten (National Ministry for Nutrition, Agriculture and Forestry), Bonn, Germany, 1994, (Europe)

- Guide for the care and use of laboratory animals, Institute of Laboratory Animal Resources, Commission on Life Sciences, National Research Council, National Academy Press, Washington D.C. 1996

## 2 Requirements for Keeping Laying-Hens in View of the Protection of Animals

Under present conditions of research-institutions both private and public, as also for commercial reasons, the scope for keeping hens in a way suited to their nature is limited, since it is seldom possible to keep them on the ground with enough space. Nonetheless the available containers (cages or coops) should enable the birds to exercise most of their natural repertoire of behaviour. This includes social contact, so at least two hens should be kept together, or in cases where hens are unavoidably kept on their own, they should be able to see and hear others. Hens which are to be immunised differently can be kept together without confusion by being chosen from differently coloured races, such as white and brown.

The various conditions under which they are kept should make possible the following forms of behaviour (Fölsch 1990, Anonymous 1994, Scharmann 1996, Schecklies 1996):

- upright standing and walking,
- scratching and looking for food,
- eating and drinking ad libitum,
- dust-bathing in a suitable substrate (like wood-granulate),
- egg-laying and withdrawal into a quiet zone,
- flapping of wings and stretching of wings and legs, and
- standing, sitting and preening on perches.

## 3 Advantages and Disadvantages of Various Ways of Keeping Laying-Hens

### Conventional ways

The following discussion of the pros and cons of various ways of keeping hens takes only three considerations into account:

- hens' natural requirements
- hygiene
- the point of keeping hens - antibody production

### Free range

**Natural requirements**

Keeping hens more or less on meadows gives them the widest scope for behaving naturally as long as the area is suitably arranged. It should, for instance, include opportunities for dust-bathing and sheltering from intense sunshine as also opportunities for stretching in all directions (see above).

**Hygiene**

In comparison with cages, an open area is not so easily cleaned once a day or even twice a week, so there is certainly a greater risk of infection with avian pathogens from wild birds or rodents. Indeed this risk may be unique to this form of hen-keeping, so more care has to be taken over the birds' health.

**Antibody production**

The immunisation of the hens is certainly as straightforward as in other ways of hen-keeping, though it may be harder to correlate eggs and hens for the sake of checking the antibody-titre processes. A possible solution might lie in using capture- or trap-nests, though the personnel would have to check the nests regularly and to document reliably the individual laying-performances. Another suitable way to identify the laying-hens might be the implantation of microchips (similarly as with pets) in combination with corresponding laying-nests. This would be rather a high technical investment but would save the bother of checking the nests regularly. As far as we know, this method has yet to be tried out, so it is not yet possible to balance the cost of the technology against the saved cost of personnel. Naturally this presupposes - not self-evidently - that the hens actually use the nests and do not lay their eggs here and there on the ground (see Jungbäck and Finkler, 1998).

### Coops

**Natural requirements** This form of hen-keeping too enables hens to satisfy most of their needs, though a distinction can be made between keeping hens out of doors, with a natural ground and buildings for shelter in winter, and keeping them indoors. In the latter case, trouble will have to be taken to make the aviary more natural. This seems to us to be no great snag. The behaviour typical of looking for food (pecking and scratching) can be made possible by strewing grain and shell-grit or short straw. Thereby, at least to some extent, feather picking can be avoided (Martin, 1986, Scharmann, 1996). It is important that the size of the ground area in the aviary be suitable for the number of hens on it. For instance the minimum for 10 hens is thought to be 4 $m^2$ (O'Donaghue, 1994). The room temperature for the hens should vary between 15-20°C at a relative humidity of 60–80%. Decisive for the acceptance of food as well as for stabilising and regulating the hormones, especially those for growth and laying, is the light. For a young hen up to 18 weeks old a day with 8-10 hours light is recommended. At the start of and during laying, this period may be extended to 12 but to not more than 16-18 hours. A light-intensity of 16 Lux should be maintained (Scholtyssek, 1987; Schwark et al. 1987, Siegmann, 1992).

To accompany veterinary measures or to introduce laying pauses, these values can be greatly reduced. For technical reasons we use a light-regime of 12 h light to 12 h darkness. High frequency light-tubes are advisable, since hens react adversely to the flickering of lower frequencies (Nuboer et al. 1992, Manser, 1996).

**Hygiene** As regards hygiene, there is a notable difference between aviarys indoors and out. For outdoor aviarys the same guidelines apply as to free-range hens. Hen-keeping indoors makes cleaning the aviarys easier, which serves to avoid infectious agents.

**Antibody production** The same applies basically as for free-range hens, but egg laying is more easily checked.

### Keeping hens in cages

**Natural requirements**

Here there is to be no discussion of various forms of cages, like batteries or so-called flat-decks, as are common in keeping hens commercially (see comments by Oester and Fröhlich 1986, 1997 and Siegmann 1992), since they seldom meet the requirements of animal protection. The cages in mind here cater on the whole to the hens' natural needs. In comparison with the above alternatives, this way of keeping hens greatly restricts their scope for behaving naturally, but on the market by now there are also kinds of cages (for instance Fa. EBECO, Castrop-Rauxel [Untiedt, 1998] or Fa. EHRET), which are generally satisfactory. They allow hens to sit on perches, to lay eggs in a peaceful zone (nest), to walk and scratch and to lie down (as above, see Figure 1a-e). For each hen there is a trough of food and a nipple-drink available. The ground area of a unit comes to 4000 $cm^2$ for two hens and the height to 80 cm, and the ground area of each unit can be split into two separate cages by means of a sliding partition. Hence this kind of cage meets the recommendations of the Swiss Office for Veterinary Medicine and also goes beyond the minimal demands set down in the 'UFAW Handbook on the Care and Management of Laboratory Animals' (1987) or the EU-Guidelines of 1990. The dimensions of the cages described above lie clearly above these minimal demands (see Table 1).

The variation in minimal demands for the height of cages is due to considering perches. According to the Guide for the care and use of laboratory animals (Anonymous 1996) the minimum room allocation for a chicken (> 3 kg b.w.) should be 3.00 square feet x 0.09 = 0.27 square meter, that amounts to ca. 0.600 $m^2$ if two chickens are kept.

**Table 1.** Comparison of the Minimal Demands in the Literature for Keeping Chickens in Cages (Relating to the Keeping of two Laying-Hens)

| | EU Guidelines (1985/90) | UFAW Handbook (1987) | Swiss Office for Vet. (1994) |
|---|---|---|---|
| Least area | 1.700 $cm^2$ | 2.400 $cm^2$ | 4.000 $cm^2$ |
| Least height | 45 cm | 55 cm | 80-100 cm |
| Food-trough (Least length) | 15 cm | 12 cm | 10 cm |

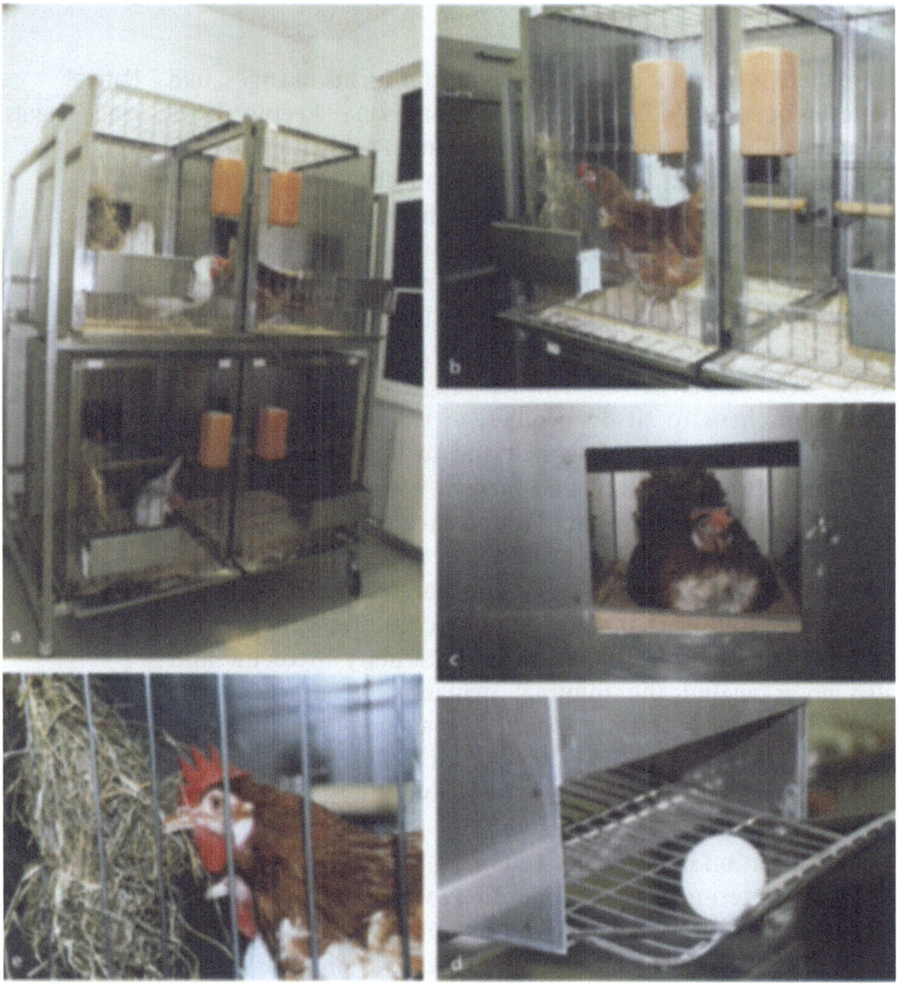

**Fig. 1.** Here is shown a cage in which hens can be kept in a way suitable to their nature, **a**) a general view of the cage consisting of four units, **b**) a front view of two units with a food-rack and nipple-drink, **c**) a view of the laying nest. The hen sits on a concave wooden frame covered in coconut-fibre matting. The frame has a hole at its deepest point, so that an egg can fall onto a mesh, **d**) a view of the rear of the lower part of the laying-nest with rack, **e**) shown is a chicken busy with a bundle of hay.

**Hygiene**

As regards hygiene, this way of keeping hens is better than others. The cleaning of cages, which are of rust-free steel, is easy, as is the provision of food (hay-rack) and water (nipple-drinker, 2 per cage-unit). The risk of infection is less than in other cases. For straw, soft-wood granulate is used, in which the birds are able to scratch. Further handouts of grain, grit and hay keep the birds busy for a long time.

**Fig. 2.** Here are shown the colours of eggs from different chicken races (white egg - white Leghorn, brown egg- brown Leghorn, grey egg - "Araucana", green egg - "Araucana").

**Antibody production** This way of keeping hens is the most suitable for the sake of registering when and where eggs are laid and of checking the titres. With a brown and a white hen in each unit, there is no difficulty in correlating the eggs and hens, as is necessary in the case of different immunisations. There is a chicken race which lays green coloured eggs, thus several combination of chicken races are possible for egg identification (see Figure 2).

## 4 SPF (Specific Pathogen Free) Way of Keeping Laying Hens

Hens, animals or animal-groups said to be pathogen free are those in which various specific micro-organisms (bacteria, viruses, fungi and so on) and their antibodies cannot be shown to be present. This may not mean that they are free from other micro-organisms. The SPF status - the freedom of the chickens and their eggs from defined micro-organisms and/or their antibodies has to be shown successfully by suitable methods (control diagnosis) at certain intervals. To maintain the SPF status certain security systems are necessary (like isolators) to keep chickens with the specified pathogens out (Prusas et al., 1992).

**Natural requirements** Basically the same applies as in the case of conventional hen-keeping in cages. Decisions about the system (with barriers or isolators) and the kinds of cage used for SPF hen-keeping are generally determined by considerations of space, security and cost rather than by considerations of animal protection.

**Hygiene** By definition, hen-keeping under SPF-conditions establishes a higher standard of hygiene than do the above-mentioned alternatives. This involves not only the isolation and sluice systems for the hens, the personnel, food, water and waste-removal but above all control diagnosis as regards the hens and eggs. For SPF hens a series of defined pathogens, as also methods suitable for checking their absence in the birds, has been stipulated (Table 2).

**Antibody production** The immunisation itself is easy enough, but it is important that the existing SPF status of the respective hens should not be impaired by the procedure adopted. Depending upon the system used, the personnel may have to use sluice-procedures on entering and leaving.

**Table 2.** Control Diagnosis for SPF Chickens (Prusas et al., 1992, Vielitz, 1997)

| Pathogens | Testing methods |
|---|---|
| chicken anaemia agent (CAA) | ELISA |
| avian adeno viruses | |
| a) serotypes 1-12 | AGPT, SN |
| b) EDS 76 | HI |
| avian encephalomyelitis | ELISA |
| avian leukosis / RSV infection | ELISA |
| avian nephritis virus | FAT |
| avian retroviruses | ELISA |
| fowl pox | CE /PM |
| infectious bronchitis (IB) | ELISA |
| infectious bursitis (IBDV) | ELISA |
| infectious laryngotracheitis | ELISA |
| influenza A-infection | AGPT / HI / ELISA |
| Marek disease | AGPT |
| Mycoplasmosis (Mg/Ms) | Aggl. |
| Newcastle disease (ND) | ELISA / HI |
| Reticuloendotheliose virus | ELISA |
| salmonella pullorum infection | Aggl. |
| other salmonella infections | BE |
| turkey rhinotrachetis virus (TRT) | ELISA |

ELISA = enzyme linked immunosorbent assay; HI = hemagglutination inhibition test; AGPT = agar-gel-precipitation test; Aggl. = agglutination test; BE = bacteriological examination; CE = clinical examination; PM = post mortem; FAT = fluorescent antibody test; SN = serum neutralisation test.

The immunisation of SPF hens is often thought to be more favourable as regards antibody production, but for conclusive proof, the antibody production of hens used to SPF keeping would have to be compared to that of hens of the same race used to conventional keeping. Moreover reactions may vary according to the respective antigens, so results able to support broad generalisations may be arrived at only at great material and financial cost. There is a study of Behn et al. (see Chapter 3 and Appendix) which demonstrates both the development of yolk-IgY concentration and the Ab-titre in chickens under SPF- versus conventional conditions. From these results and our own experience, we believe that the immunisation of SPF hens has no marked advantage over that of hens kept conventionally. But the discussion about this may still be open.

**Summary** A comparison of the various forms of hen-keeping suggests that the greater the scope for hens' natural behaviour the harder it is to maintain hygiene and to correlate hens and eggs. The relation is nearly reciprocal.

## 5 Egg-Laying Performance of Chickens Kept in Cages After Immunisation

Our experience suggests that hens kept in cages are very robust as regards egg-laying. Naturally the performance may vary from race to race. We use hens belonging to or derived from the following races: White Leghorns and/or medium-weight brown hens (Rhodeländer, New Hampshire). An essential question is how continuous the laying performance is and how it is affected by immunisation. Basically the laying performance does not vary according to how the hens are kept (Fölsch, 1990; Horne and Nieker, 1997). Our experience suggests that the laying performance of hens in conventional cages as above is somewhat better in comparison with the performance of hens in SPF cages. According to the literature, egg-laying is normally not interrupted by immunisation (Schade et al., 1996, see Figure 4, 5). The performance may be slowed down, but this is due mostly to wounds, infections, social stress, moulting, a change of diet and so on. There is evidence that the nature of the antigen may affect the laying performance more strongly than the use, for instance, of Freund's adjuvant (Schade et al., 1994). According to investigations by Erhard et al. (1996), hens may react briefly to booster-immunisation with the use of Freund's complete adjuvant (FCA) by laying fewer eggs, but in the following period they very nearly make up for this in comparison to control hens (see Figure 3).

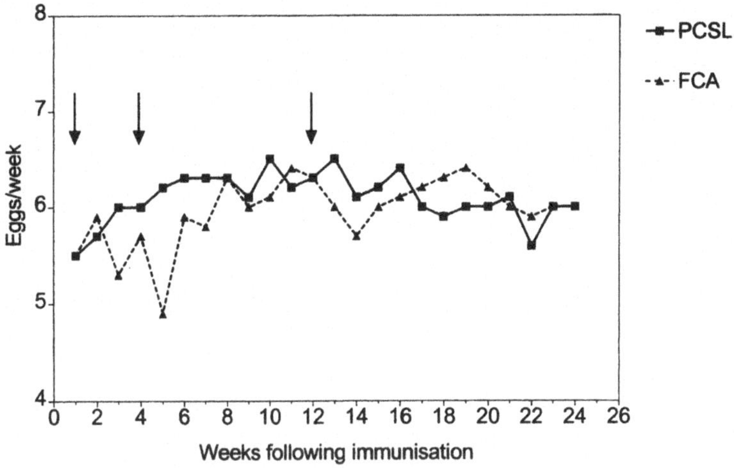


**Fig. 3.** Hens' average laying performance (n = 6, conventional keeping) after immunisation with FCA or PCSL (see Chapter 3). The arrows show the times of immunisation and boosting. FCA, but not PCSL, is seen to influence the laying performance for a while.

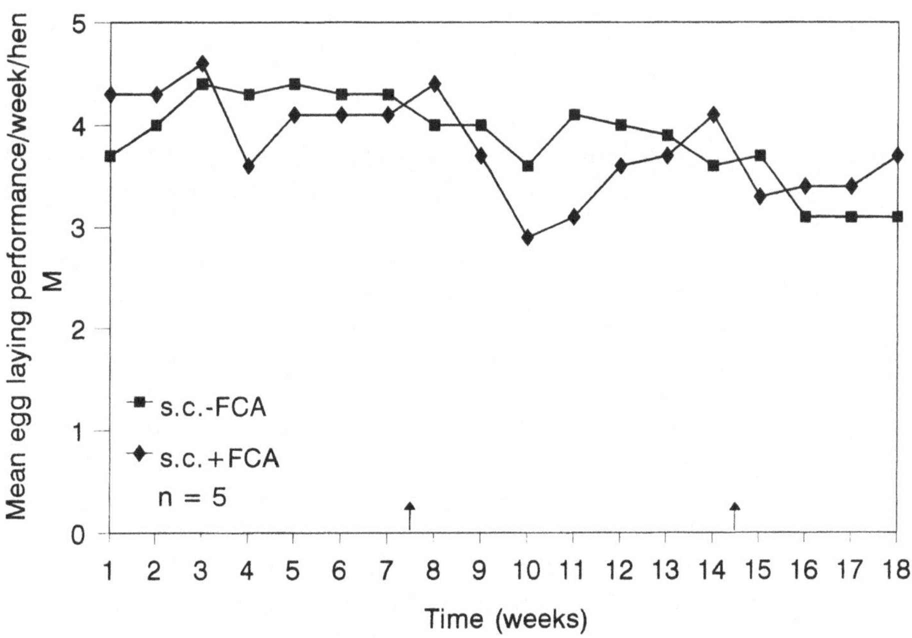


**Fig. 4.** Hens' average laying performance (n = 5) under SPF conditions in a flat deck (see Schade et al. 1994). There is a comparison between the laying performance after subcutaneous (s.c.) immunisation (arrow) with and without FCA. Here too can be seen a transient change in the laying performance after FCA. The laying performance is less than in the case of Figure 3.

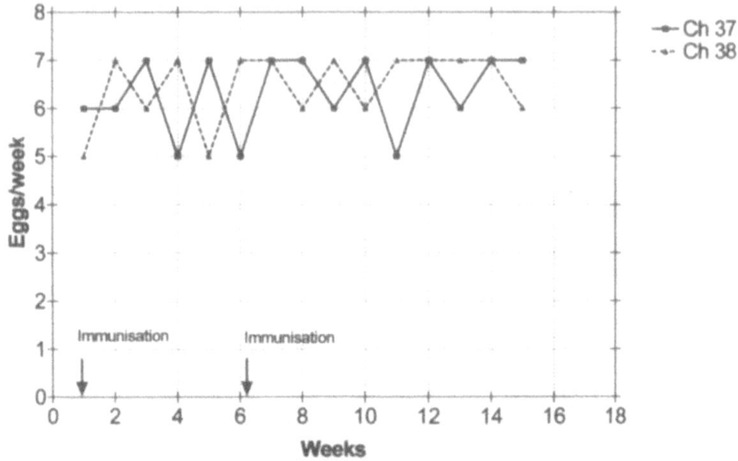


**Fig. 5.** The weekly laying performance of two hens kept in the special cages described. They are immunised with FCA or FIA. The weekly laying performance fluctuates between 5 and 7 eggs per week. In this case the use of FCA does not notably affect the laying performance.

## 6 Feeding

The hens can be fed easily by the use of racks attached to the outside of the cages and be given a defined meal for laying-hens in the form of flour, pellets or granulate ad libitum.

Moreover the hens are also given corn-food (like wheat) or shell-grit. The additional food is strewn on the cleaned floor on which straw has been spread for them to scratch in. A wire mesh with large holes spread over the floor prevents too much straw from being tossed out. To prevent hens from pecking pointlessly as a kind of behavioural stereotype (Martin, 1986; Oester and Fröhlich, 1986; Scharmann, 1996) which may also take the form of feather-picking, we hang bundles of hay in the cages, so the birds can practice various forms of behaviour - like pecking, plucking and tearing - associated with taking up plant-materials. This also keeps them busy for longer periods.

**Table 3.** Mixed Meal for Laying-hens (DLG-Standard, 1987, Siegman, 1992)

| Ingredient | Amount (minimum pay) |
|---|---|
| raw protein | 15 % |
| methionine | 0.30 % |
| methionine / cystine | 0.85 % |
| lysin | 0.63 % |
| sugar | 12 % |
| calcium | 3.20-4.00 % |
| phosphorus | 0.50-0.80 % |
| sodium | 0.12-0.25 % |
| | |
| manganese | 40 mg/kg |
| zinc | 60 mg/kg |
| vitamin A | 8000 IU |
| vitamin D5 | 1000 IU |
| vitamin E | 10 mg/kg |
| vitamin K5 | 1 mg/kg |
| pantothenic acid | 10 mg/kg |
| thiamine | 2 mg/kg |
| riboflavin (B2) | 2.5 mg/kg |
| nicotine acid | 30 mg/kg |
| vitamin B6 | 3 mg/kg |
| choline | 1000 mg/kg |
| folic acid | 0.7 mg/kg |
| energy content (ME) | 10.6 MJ |

## 7 Health Protection

In being kept as poultry, hens are immunised against many pathogens, to protect them from various poultry ailments and animal infections. Without these extensive immune-prophylactic measures, modern poultry-keeping would be unthinkable. Hens used for producing specific antibodies are generally kept for this purpose from youth on or from the age of egg-laying, and by this time they have adequate immunity towards a range of possible pathogens. SPF hens (see above) are exceptions in this respect. When hens are being kept for laboratory purposes, immune-prophylactic measures are needed according to what seems to be the risk of infection and the respective epidemiological relations. In most countries it is mandatory to immunise them against Newcastle disease and in many countries against fowl plague. In the case of hens kept for producing antibodies, the necessary immunisations should be arranged with the veterinary authorities responsible.

The hens should be checked regularly for human pathogens and causal agents of zoonosis, especially for salmonella spp. and Chlamydia infections.

According to where the birds are from and how they are being kept, they should be checked for ecto- and endoparasites. Addition of anticoccidial agents to the food is recommended, especially for hens kept on the ground.

## 8 Suppliers

### Cages for chicken keeping

E. Becker & Co GmbH (EBECO)
Hermannstr. 2-8
D-44579 Castrop-Rauxel
Tel.: 02305-97 30 40
FAX: 02305-97 30 444

EHRET GmbH & Co KG
PB 1230
D-79302 Emmendingen
Tel.: 07641-92 65 0
FAX: 07641-4 79 72

## References

Anonymous (1988) Publication of the "Ausschuss für Ausbildung und tiergerechte Haltung" der GV-SOLAS concerning "Planung und Struktur von Versuchstierbereichen tierexperimentell tätiger Institutionen" Nr. 1, 4. Reviewed Edition, May 1998.

Anonymous (1985/1990) European Convention for the Protection of Vertebrate Animals used for Experimental and other Scientific Purposes. Council of Europe, Strasbourg 31.05.1985, Table 13, Guidelines for Caging Birds. (see also Bundesgesetzblatt Teil II Nr.46, 15.12.1990).

Anonymous (1994) Richtlinien betreffend Käfigabmessungen und -einrichtungen für die Haltung von Haushühnern zu Versuchszwecken, Bundesamt für Veterinärwesen, 3097 Liebefeld-Bern, 17. Jan. 1994 (2) d Vo/Oe/Fr/St/re-800.116-2.04.

Anonymous (1996) Guide for the care and use of laboratory animals, Institute of Laboratory Animal Resources, Commission on Life Sciences, National Research Council, National Academy Press, Washington D.C. 1996.

Bundesgesetzblatt Jahrgang 1998 Teil I Nr. (1998) Bekanntmachung der Neufassung des Tierschutzgesetzes vom 25. Mai 1998 30, ausgegeben zu Bonn am 29. Mai 1998.

Erhard MH, Hofmann A, Stangassinger A, Lösch U (1996) Einfluß verschiedener Adjuvantien auf die Legeleistung bei der Immunisierung von Hühnern. In: Schöffl H, Spielmann H, Tritthart HA (eds) Ersatz- und Ergänzungsmethoden zu Tierversuchen. Springer, Wien New York, p. 416.

Fölsch DW (1990) Ethological aspects of the behaviour of hens in relation to different housing systems. Dtsch. Tierärztl. Wschr. 97:228-230

Horne PLM van, NiekerkTGCM van (1997) Produktivität, Gesundheit und Wirtschaftlichkeit von Legehennen in Volieren- und Bodenhaltung. Lohmann Information 3, 1997:13-14

Jungbäck C, Finkler H (1998) Observations on the effects of the flooring used for laying hens. Tierärztl. Umschau 53:303-308

Manser CE (1996) Effects of lighting on the welfare of domestic poultry: A review. Animal Welfare 5:341-360

Martin G (1986) Die Pickaktivität von Hühnern als Kriterium für tiergerechte Fütterungs- und Haltungsbedingungen. In: Aktuelle Arbeiten zur artgemäßen Tierhaltung, KTBL-Schrift 311,. Hiltrup, Landwirtschaftsverlag, Münster, pp. 116-133

Nuboer JFW et al. (1992) Artificial lighting in poultry jouses: Do hens perceive the modulation of fluorescent lamps as flicker ? British Poultry Science 33:123-133

Oester H, Fröhlich E (1986) Die Beurteilung der Tiergerechtheit der neuen Haltungssysteme für Legehennen im Rahmen der Tierschutzgesetzgebung. Schwei. Arch. Tierheilk. 128:521-534

Oester H, Fröhlich E (1997) Die Haltung von Haushühnern zu Versuchszwecken. In: Das Buch vom Tierschutz, Ed.: Enke Verlag Stuttgart, pp. 397-400

O'Donoghue PN (1994) The accommodation of laboratory animals in accordance with animal welfare requirements, Proceedings of an International Workshop held at the Bundesgesundheitsamt, Berlin, 17.-19. Mai 1993, Bundesministerium für Ernährung, Landwirtschaft und Forsten, Bonn, Germany, pp. 53-58

Prusas E, Prusas Ch, Heider G (1992) Spezifiziert pathogenfreie Hühnerzucht. In: Heider G, Monreal G (eds) Krankheiten des Wirtschaftsgeflügels. Gustav Fischer Verlag Jena Stuttgart, pp.

Schade R, Bürger W, Schöneberg T, Schniering A, Schwarzkopf C, Hlinak A, Kobilke H (1994) Avian egg yolk antibodies. The egg laying capacity of hens following immunization with antigens of different kind and origin and the efficiency of egg yolk antibodies in comparison to mammalian antibodies. ALTEX 11:75-84

Schade R, Rieger A, Hlinak A (1996) Möglichkeiten und Grenzen der Gewinnung polyklonaler Antikörper aus dem Hühnerei. IgY-Technologie. Tierlaboratorium 19:41-59

Scharmann W (1996) Tiergerechte Haltung von Legehennen unter Laborbedingungen. Tierlaboratorium 19:67-71

Schecklies E, Polyklonale Antikörper. Eine Einführung in die Theorie und Praxis der Antikörperherstellung. VCH Verlagsgesellschaft mbH, Weinheim New York Basel Cambridge Tokyo, 1996.

Scholtyssek S (1987) Geflügel. Verlag Eugen Ulmer, Stuttgart, pp.

Schwark HJ, Mazanowski A, Peter V (1987) Internationales Handbuch der Tierproduktion. Geflügel. Deutscher Landwirtschaftsverlag, Berlin, 1987

Siegmann O (1992) Propädeutik. In: Heider G, Monreal G (eds) Krankheiten des Wirtschaftsgeflügels. Gustav Fischer Verlag, Jena Stuttgart, pp.

UFAW (1987) The domestic fowl and turkey. In: Handbook on the care and management of laboratory animals (Ed.: Poole TB). Longman Scientific & Technical, 6th Edition, p.645.

Untiedt H (1997) Tierschutzgerechte Käfighaltung von Legehennen für die Produktion von Ei-Antikörpern. Der Tierschutzbeauftragte 6:192-194

Vielitz E (1997) Geflügelgesundheit- Rückblick auf vier Jahrzehnte. Lohmann Information 3:25-30

# Immunisation

CHRISTINE SCHWARZKOPF, CHRISTIAN STAAK, IRENE BEHN, and MICHAEL ERHARD

## Introduction

In this chapter, experiences in immunising hens are described. These experiences are based on the results of several studies involving various antigens as well as various adjuvants. Results of these studies can be generalised only within wide limits. The reason is, that the immune response is affected by three variables: Antigen, adjuvant, and the immunised animal. The most important of these is the individual animal, since even without an adjuvant an immune response can be achieved in some cases. The results presented here may be of help to users in enabling them to select optimal conditions for immunisation of chickens.

### Properties of antigens influencing the development of specific Ab

As already mentioned above, an immune response is essentially influenced by the quality and amount of the antigen used.

Already in 1945 Landsteiner pointed out that there exists a relation between the molecular weight and immunogenicity.

✉ Christine Schwarzkopf, Zentrales Institut des Sanitätsdienstes der Bundeswehr Koblenz, Laborabt. II (Vet.Med), Generaloberst-Beck-Str. 1f, Mainz, 55129, Germany (*phone* +49-6131-56-3402; *fax* +49-6131-56-3409)
Christian Staak, Bundesinstitut für gesundheitlichen Verbraucherschutz und Veterinärmedizin, Diedersdorfer Weg 1, Berlin, 12277, Germany
Irene Behn, Universität Leipzig, Fakultät für Biowissenschaften, Pharmazie und Psychologie, Talstr. 33, Leipzig, 04103, Germany
Michael Erhard, Universität Leipzig, Veterinärmedizinische Fakultät, Veterinär-Physiologisches Institut, An den Tierkliniken 7, Leipzig, 04103, Germany

Haptens of low molecular weight are not immunogenic unless coupled to a carrier of higher molecular weight. For booster-injections, the same carrier-bound hapten has to be used to reach an adequate increase in titre. A co-operation between carrier-specific T-helper-cells and hapten-specific B-lymphocytes positively influences hapten-specific antibody production (Snippe and Kamp, 1975). In the case of multiple immunisations, the interference of antigens and a reciprocal influence on the immune responses should be taken into account. For instance, the humoral response to an immunisation could decrease the intensity of a cell-mediated response to a further immunisation at the same time. The response to a single antigen can be greater than an immune response following an application of two competing antigens (Dunnington, 1992).

**Natural and artifical protein antigens** As with mammals, immunisation of hens against recombinant antigens is possible. The criterium for an antigen as a good immunogen is the sequence of amino-acids to be recognised by the hens immune system as a foreign one. That applies to antigens of natural as well as of synthetic origin.

Moreover, molecules not arising naturally can be synthesised. Hence B-cell epitopes and T-cell epitopes can be united in a single antigen and then used for immunisation. The immunisation of hens with proteins results normally in a good Ab-production. The minimum molecular weight necessary to obtain a sufficient immune response is similar in both, mammals and chickens and ranges from 5 to 10 kD.

**Peptides (haptens)** It is often discussed, that hens only react moderately, if at all, to immunisation with peptides. This opinion is largely unfounded. If haptens are coupled to immunogenic molecules (carriers), antibodies can be raised also in hens. The immunogenicity of a peptide depends largely on the sequence of amino acids, on the method of coupling to a carrier and on the carrier itself. Moreover the choice of adjuvant is of relative importance. The quality of an antibody also depends essentially on the fact as to whether the synthetic peptide represents cryptic or accessible structures of the original protein (Boersma et al., 1993). This may also determine whether or not the antibody obtained reacts not only with the corresponding peptide sequence but also with the native protein.

Specific antibodies against peptides can be raised in chickens which are suitable for use in various test-systems. For instance, antibodies can be produced against various sequences of the neuropeptide cholecystokinin octapeptide. They can be used for immunohistochemical studies as well as for quantitative measurements (Schade et al, 1996). Concentrations of 100 - 250 μg peptide/immunisation should be applicated.

**Carbohydrates**

Like peptides, carbohydrates have a reduced immunogenicity dependent on their structure (pure or in combination with proteins: glycoproteins, lipoproteins). They induce a short T-cell-independent immune response. Coupled to carrier proteins, they induce a T-cell-dependent response and hence an increase in the production of antibodies. Due to a limited spectrum of antigenic determinants and a wide distribution in living systems carbohydrates are involved in a lot of immunological crossreactions. Chickens were successfully immunised against glycoproteins (see Chapter 8).

**Nucleic acids, nucleoproteins**

Since the immunogenicity of nucleic acids is very low, coupling to a carrier is necessary. Experiences concerning the immunisation of chickens with nucleic acids, nucleoproteins see (Chapter 8).

### Influence of the route of application and the amount of antigen on development of specific Ab

The amount of antigen used is not irrelevant to the immune response. It seems that there are optimal concentrations of respective antigens (Schecklies 1996, own data). For instance, 0.1 or 0.5 mg of mouse-IgG elicits a maximal antibody titre, whereas an increase to 1 mg reduces the immune response (Behn et al. 1996, see Chapter 8). Comparable data were shown by Sasse et al. (1998) following the immunisation of chickens with endotoxins. Optimal doses of antigens range between 10 μg and 1 mg (Mahn 1998).

Often in the case of booster-immunisations, a smaller amount of antigen is used than for primary immunisation. The idea is that in this way the highly affine B-cell receptors bind the antigen and mainly these B-cell clones are stimulated.

This is meant to lead to antibodies with high affinity. According to general experience, the greater the purity of an antigen for immunisation, the lower is the immunogenic potency (Hudson and Hay, 1991).

Antigen and adjuvant are injected by the intramusclar (i.m.) or subcutaneous (s.c.) route. Other forms of application have been described (see Schade et al. 1992, Leenars et al. 1999).

### Influence of hens' keeping conditions on development of specific Ab

Judging by results so far, the various ways of keeping hens (as detailed in Chapter 2) do not essentially influence the nature of an immune response, unless, for instance, the animals suffer from an infection due to contact with an infectious environment.

The existence of an individual background repertoire of various antibody-specificities is a phenomenon shared with other species and which in most cases is not disturbing. Otherwise the specific antibodies could be purified and concentrated by means of affinity chromatography. There are a few cases where the immunisation of SPF-chickens is advisable but this kind of antibody production is normally not necessary and, for economic reasons, will remain an exception.

The results of comparative investigations of antibody titre development in hens kept conventionally or under SPF conditions showed no essential differences (see Chapter 8). Very interesting is the fact that the concentration of total IgY rises much more after immunisation in SPF-hens than in hens kept conventionally (see Table 1), whereas the development of Ab-titre is similar in both cases. The IgY content in SPF-hens reaches nearly the same level as in hens conventionally kept but SPF animals start at much lower levels before immunisation.

One can suggest that due to the adjuvant/antigen challenge the immune system of the SPF-hens is not only stimulated to produce antigen specific IgY but also unspecific antibodies and that the rise in the titre of specific IgY in hens kept conventionally takes place on account of other IgY specificities.

**Table 1.** Comparison of the Overall Yield of IgY (mg/egg yolk) From the Yolks of Chickens Kept Conventionally and Kept Under SPF Conditions, Respectively. Days of Immunisation: 0 and 28, Days of Testing: 5, 33 and 100

| | Conventional-keeping | | | SPF-keeping | | |
|---|---|---|---|---|---|---|
| Days | 5 | 33 | 100 | 5 | 33 | 100 |
| Mean value (n=10) | 39,6 | 58,8 | 61,4 | 18,0 | 38,7 | 49,4 |
| SD | 6,6 | 11,2 | 12,0 | 6,4 | 7,1 | 10,8 |

The chickens (n=10) were immunised with different immunoglobulin isotypes.

## Adjuvants - history and theory of the effective mechanisms

The production of polyclonal antibodies is supported by the use of B-cell stimulating immune modulators (adjuvants).

Although this stimulation takes place independently of the kind of antigen and is thus non-specific, it improves the quantity and specificity of an immune response (Freund, 1937, 1944; Glenny et al., 1926; Kaeberle, 1986; Roitt et al., 1987; Thein, 1988). The exact way in which they work is still not wholly understood. They promote both the humoral and the cellular immune response and induce an immunological memory. A certain long-term effect and thus a protracted release of the antigen contribute to the effect as does a local inflammation after immunisation.

Adjuvants can wholly or partly influence the composition of the immunoglobulin isotypes created during the immune response (Mallon et al., 1991). This can apply to both, T-cell dependent B-cell stimulation (muramyldipeptide) as well as to T-cell independent B-cell responses (especially LPS). The T-cell stimulator muramyldipeptide (MDP) is a constituent of the cell wall of mycobacteria with a strong immunogenic potential (Ellouz et al., 1974). On the other hand B-cells are activated strongly by lipopolysaccharide (Johnson et al., 1956; Azuma et al., 1976; Sugimoto et al., 1978; Mayr et al., 1984; Sprick-Sanjosé Messing, 1990; Lübke, 1990). CD14 has been identified in its function as an LPS-binding protein (Schütt et al., 1992). CD14 belongs to the phospholipid-inositol proteins and is expressed on monocytes and macrophages. Ligand-binding to CD14 activates the phospholipase $A_2$ (and also the arachidonic acid metabolism), so contributing to the initiation of inflammatory mechanisms. Shedded CD14 in interstitial tissues partly neutralises LPS.

The concept of the adjuvant was devised by Ramon in 1925. There was early use of starch, agar and saponins as a basis for adjuvants. Even nowadays saponins are part of many veterinary vaccines and are extracted from the bark of the south American tree *Quillaia saponaria.*

In 1926/31, as Glenny et al. put forward the theory of the deposit effect of adjuvants, they used mainly aluminium hydroxide [$Al(OH)_3$], whose effect as an immune stimulant depends on the amount of antigen absorbed. The optimal ratio between the antigen and $Al(OH)_3$ has to be found out separately in each case (Bomford, 1985). The protracted release of the antigen from the complex prolongs its phase of presentation to immune-competent cells (Kellner, 1990). But even more than in the case of oil-in-water emulsions, the stability and hence the deposit effect of the antigen-$Al(OH)_3$ complex is of limited duration. Subcutaneous application leads to the formation of aluminium granulomas (Nail et al., 1976; WHO, 1976).

The adjuvants most commonly used in animal experiments are applied as water-in-oil emulsions. In these emulsions the antigen is in the watery phase, as for instance with Freund's complete adjuvants (FCA) (Finger et al., 1964; Herbert, 1974; Sanchez et al., 1980; Lösch et al., 1986; Thein, 1988).

A grave disadvantage of the use of adjuvants proved to be local and systematic side-effects, so intensive research into possible alternatives began. The following side-effects were to be lessened or eliminated:

- Sterile abscesses and painful granulomas tending towards ulceration, which may lead to systematic fistulas with metastatic spreading (Steiner et al., 1960; Schmidt et al. 1996, Wanke et al. 1996; Bennett et al., 1992),
- the facultative carcinogenic effect (Potter, 1977; Broderson, 1989),
- the risk of allergy in the case of repeated use (Beebe et al., 1972),

- falsely positive results with tuberculin tests after the use of FCA (Mayr et al., 1984),
- an adverse effect on meat quality if used on farm animals (Allison et al., 1986).

A further possible side-effect of adjuvants is the disruption of tolerance towards autoantigens present at the same time as the foreign ones (Nakashima et al., 1979; Paterson, 1966; Broderson, 1989; Amyx, 1987). The animal model of arthritis induced by adjuvants may be taken as an example (Tal and Laufer, 1969). For these reasons, many adjuvants cannot be used on humans (see section "side effects").

### Commercially available adjuvants

Beside the adjuvants investigated in chickens and described below, the report of the ECVAM workshop 35 "The production of polyclonal antibodies in laboratory animals" (Leenaars et al. 1999) gives more detailed information on other adjuvants and their immune stimulating effects above all in the use in mammals.

### Water in oil formulations

**FCA/FIA**

After Glenny et al. (1926, 1931) had set up the theory of the long-term effect of adjuvants, Freund et al. (1937, 1942, 1951) developed the theory further. They used water-in-oil emulsions made up of an antigen in a watery solution and paraffin-oil as well as inactivated mycobacteria (Freund's complete adjuvant, FCA) for immunisation. In the emulsion, the antigen is resistant to dispersion and, as a stable depot, effects a prolonged antigen stimulation (Bomford, 1985; Schmidt et al. 1996, Wanke et al. 1996). Mineral oils likewise tend to create sterile abscesses and hence increase the depot effect (Thein, 1988). By adding heat-inactivated mycobacteria the adjuvant-properties can be additionally improved (Finger, 1964; Mayr et al., 1984). A cause of this is the activation of macrophages leading to the release of antigen non-specific factors (cytokines), which strengthen the immune response.

FCA is taken as the so called "gold standard" and is the adjuvant most often used for immunisation. It is an extraordinary potent immune stimulant, and owing to the good immune response to even weak antigens, its often undesirable side-effects are tolerated.

Freund's incomplete adjuvant (FIA) contains no preparations from mycobacteria.

**Adjuvant 65** This is the only adjuvant in which a stable emulsion results from using oils from plants: peanut-oil with aluminium monosterate as a stabiliser (Hilleman, 1967). Like FCA, FIA and adjuvant 65 need at least 50% oil for a stable emulsion (Mallon et al., 1991; Bennett et al., 1992).

**Specol** Specol was first described by Bokhout et al. in 1981 and is made up of a mineral oil and the emulgators Span 85 and Tween 85. For immunisation a water-in-oil emulsion is made with the antigen in the watery phase. The emulsion remains stable for a lengthy period: one year at 4°C and three months at 37°C. (Boersma et al., 1992). Ready to use, the viscosity of the emulsion is low.

**TiterMax** TiterMax is likewise a new synthetically made adjuvant, which was developed to reduce the undesired side-effects as described with FCA. In TiterMax the advantages of a stable, water-in-oil emulsion are combined with the potent adjuvant activity of a synthetic polymer made up of hydrophobe polyoxypropylene and hydrophile polyoxyethylene. A free polymer and a polymer bonded to silicium dioxide form, together with a degradable oil (squalene), a stable emulsion.

### Oil in water formulations

**ABM-system (comparable to the ribi-system)** Another new adjuvant formulation is the ABM-adjuvant system. It forms stable oil-in-water emulsions of low viscositiy, requiring only 2% oil (squalene). In chickens there are no abcesses due to immunisation. The components used in this adjuvant system are trehalosedimycolate (TDM), cell-wall skeleton (CWS), monophosphoryl lipid A (MPL) as well as inactivated pathogenic bacteria and Tween 80 as an auxiliary.

Gerbu

Glycopeptides from the cell walls of mycobacteria have an immune stimulatory effect, for instance N-acetylglucosamine-MDP (GMDP) used in the Gerbu adjuvant. This also contains lipids with quaternary ammonium bonds like DDA (dimethyl dioctadecyclammoniumchloride). They bind not only the antigen but also the GMDP and are thus phagocytised. Further components of the adjuvant are L-proline, lecithin and Tween 80.

Lipopeptide PCSL

In various animals, an immune stimulatory effect has been shown for synthetic lipopeptides together with a whole series of antigens (Wiesmüller et al., 1983; Kellner 1990, Kellner et al. 1992, Mittenbühler et al. 1997). Among the lipopeptides so far investigated, the most effective has proven to be lipohexapeptide $Pam_3Cys$-Ser-$(Lys)_4$-OH (PCSL). No undesired side-effects have been observed (Erhard et al. 1994, Hoffmann et al. 1996, Erhard et al. 1997, Mahn 1998).

## New ways of immunisation

ISCOMs

In ISCOMs (immune-stimulating complexes), antigens are combined with micells of the quil-A-saponin, making a complex 30-40 nm in diameter. In the complexes, enriched antigen concentrations of 1 µg induce a cellular and, a humoral immune response as well. ISCOM-associated antigens are not deposited at the place of application but were transported further through the lymph system (Morein et al., 1993). The immune reaction against the membrane proteins of viruses (rabies, parainfluenza) has proven to be about ten times greater than in tests with immunisation against the proteins alone (Jilg, 1996).

Liposomes

Liposomes are artificial products constructed from natural products and are suitable for immunisations against membrane proteins. Bilayer vesicles among multi- or unilamellar vesicles are nearest to the natural cell membrane. Here, the antigen may be incorporated in the water or lipid phase of the membrane or may be coupled to the surface of the liposome (Alving 1997).

DNA-immunisation

Plasmid-DNA, which contains the gene for a respective immunogen, passes after intramuscular immunisation even into muscle cells, where the corresponding protein is expressed. This leads mostly to a T-cell dependent immune response with the production of antibodies (Jilg 1996). Possible side effects are the induction of auto-immune phenomena and the activation of oncogenes after integration of the alien DNA into the body's own cells. This DNA-immunisation does not lead to a passage of genetic material into the placenta.

**MAP (multiple antigen peptide)** Immunisation with a hapten-protein carrier complex often induces Ab against the highly immunogenic carrier, too. These specific Ab are the reason for unexpected cross reactions. To avoid such problems peptides can be coupled to trifunctional amino acids as e.g. lysine to form an immunogenic matrix. The lysine residues themselves (as the "core matrix") are relatively non-immunogenic. The greater the number of lysine residues the greater the number of coupled peptides (Tam 1988). We have good experience with a synthetic construction of resine - polyethylenglycol (PEG-as a spacer)-$(Lys)_4$-CCK8. The immunisation of chickens with this antigen, emulsified with FCA led to specific Abs able to visualise pyramidal neurons in the cortex of rat brain (see Chapter 5).

### Experimental investigations into the effect of various antigen/adjuvant combinations on development of specific antibodies in hens

In searching for alternatives to FCA, one should be concerned not only about tolerance but also about immunological properties.

Until now only a small number of adjuvants were compared in hens. There has still been no systematic presentation of the effects of various adjuvants, given various schemes of immunisation and various antigens. In 1991, Erhard et al. compared the immune stimulating effect of FCA, DPLA (diphosphoryl-lipid A) and aluminium hydroxide on various animal species, including hens. The substance chosen as immunogen was the hapten-carrier-complex formed of an organic phosphoric acid ester (hapten) and human serumalbumin (carrier). Thereby DPLA and $Al(OH)_3$ proved to have little adjuvant effect.

A detailed review is given by Leenars (1994, 1995), who experimentally compared several potent adjuvants (FCA, Titer-Max, Specol, ABM=Ribi) and ISCOMs in combination with various antigens (myelin basic protein of MW 18 kDa (MBP), mycoplasms, high-molecular Ag) in rabbits and mice.

Due to the phylogenesis there are structural and functional differences between the avian and mammalian immune systems. Therefore, the effects of adjuvants may not always be the same in hens and mammals. Schwarzkopf et al. 1997 investigated and compared the development of titre and avidity after immunising hens with a strongly immunogenic antigen (xenogenic IgG, see Subprotocol 1), using the adjuvants FCA, Specol, TiterMax, ABM-N, ABM-S, the adjuvant of the veterinary vaccine Neoparasec®, the Gerbu adjuvant and the lipopeptide PCLS. See below in this chapter for the results.

**Table 2.** Commercially Available Adjuvants Investigated and Compared on Chickens

| FCA | Difco, Detroit, USA |
|---|---|
| Specol | Instituut voor Diergezondheid (ID-DLO), Lelystad, The Netherlands |
| TiterMax | Serva, Heidelberg, Germany |
| ABM-N, ABM-S | Linaris, Bettingen, Germany |
| Adjuvant of the veterinary vaccine Neoparasec® | Rhone Merieux, Laupheim, Germany |
| Gerbu adjuvant | Gerbu Biotechnik, Gaiberg, Germany |
| Lipopeptide PCSL | Boehringer, Mannheim, Germany |

**Table 3.** Lipophile Components of Adjuvants

| Adjuvant | Lipophile component |
|---|---|
| FCA | non-metabolisable paraffin oil |
| Specol | non-metabolisable paraffins and cycloparaffins |
| TiterMax | metabolisable oil: squalen |
| ABM-N, ABM-S | metabolisable oil: 2% squalen |
| Neoparasec® | non-metabolisable paraffin oil |
| Gerbu adjuvant | DDA |
| Lipopeptide PCSL | fatty acids |

In further investigations the most effective adjuvants were combined for immunisation with a strong and a weak immunogen as well as with a corpuscular antigen. The long-term effect of the immunisation on the respective humoral immune responses was also investigated.

Both the effectiveness and the side-effects of adjuvants seem to depend on the chemical structure of their lipophile components, so before each immunisation a choice must be made between an adjuvant with or without lipids.

## Subprotocol 1
## Development of Antibody-Titre and Avidity Following an Immunisation With a Strongly Immunogenic Antigen (human IgG) in Combination With Various Adjuvants

For a general introduction see above. Here are given definitions concerning Ab-titre and avidity.

- As with mammals the antibody titre development varies with the adjuvant used. The criterion for evaluating the effect of the adjuvant is taken to be the immune response after application of the antigen without an adjuvant.
- Avidity (functional affinity) is the force with which a multivalent antibody binds a multivalent antigen, whereas (intrinsic) affinity is the force with which a single antibody binding-site binds a single antigen determinant (Roitt et al. 1991).

### Outline

The following tables show both the amounts of the antigen and adjuvant but also the scheme of immunisation. The emulsion is injected subcutaneously.

**Table 4.** Antigen/Adjuvant per Application

| Adjuvant | FCA | Specol | TiterMax | ABM N/ ABM S | Neoparasec® | Gerbu | PCSL | NaCl |
|---|---|---|---|---|---|---|---|---|
| Volume | 0.5 ml | 0.5 ml | 0.1 ml | 1.0 ml | 0.5 ml | 0.1 ml | 0.1ml | 0.5 ml |
| Antigen IgG | | | | | | | | |
| primary | 1.0 mg | 1.0 mg | 1.0 mg | 1.0 mg | 1.0 mg | 1.0 mg | 1.0 mg | 1.0 mg |
| boost | 0.5 mg | 0.5 mg | 0.5 mg | 0.5 mg | 0.5 mg | 0.5 mg | 0.5 mg | 0.5 mg |
| Route of application: subcutaneous (volume of adjuvant/antigen solution 1.0 ml) | | | | | | | | |

## Materials

**Equipment**

- syringes (antigen injection)
- multi-channel pipettes
- ELISA-reader
- microtitre-plates
- centrifuge
- refrigerator
- incubator

**Reagents, solutions**

See Chapter 5 for ELISA and Chapter 4 for IgY-extraction. Adjuvants according to Table 4.

## Procedure

1. The chickens were immunised with human IgG according to the schedule in Table 5.
2. The IgY was extracted by the water dilution method as described in Chapter 4.
3. The development of Ab-titre was consecutive determined by a respective ELISA as described in Chapter 5.
4. Isolated IgY was concentrated to 2 ml/egg and used for the avidity measurement according to the slightly modified

method of MacDonald R.A. et al. (1985). The antibody binding capacity was determined in the presence of rising concentrations of ammonium thiocyanate. The molarity of the ammonium thiocyanate solution necessary to dissociate 50% of the formed antigen-antibody-complex (reduction of the initial light absorption in ELISA by 50%) is taken as the avidity index (AI).

**Table 5.** Immunisation Schedule

| | |
|---|---|
| I. Antigen/Adjuvant (Water-in-Oil Formulation) | |
| FCA | Day: 0 / 21 |
| Specol | Day: 0 / 21 |
| TiterMax | Day: 0 / 21 |
| II. Antigen/Adjuvant (Oil-in-Water Formulation) | |
| ABM-N / ABM-S | Day: 0 / 21 / 35 / 49 |
| Neoparasec® | Day: 0 / 21 / 35 / 49 |
| Gerbu adjuvant | Day: 0 / 21 |
| Lipopetide: PCSL | Day: 0 / 21 / 35 / 49 |
| III. Control | |
| NaCl | Day: 0 / 21 / 35 / 49 |

## Results

### Antibody titre development

Until now, FCA has proved to be the most potent adjuvant.

As can be seen from Figure 1, only the effects of Specol and, to some extent, of TiterMax come up to the effect of FCA after two subcutaneous applications in the given scheme.

The effects of the adjuvants ABM-N/S, Gerbu, PCLS and Neoparasec® (data not shown) hardly differ from those of the control without application of adjuvants. They induce no useful titres for use in diagnostic systems. (In the diagram the useful titre is shown by a horizontal line.)

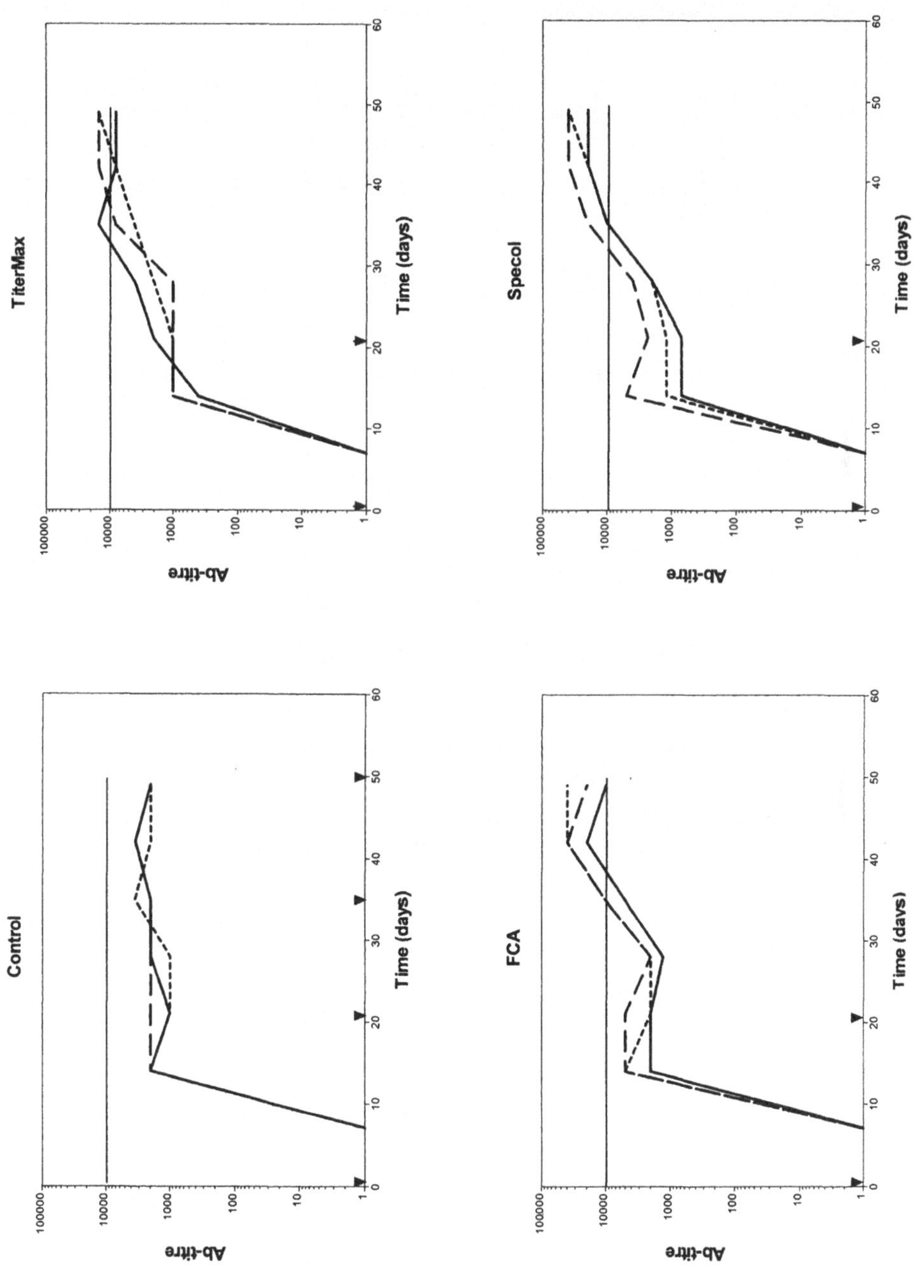


**Fig. 1.** A comparison of the efficacy of various adjuvants on antibody titre-development after immunisation with human IgG. Titres taken as valid have levels of dilution (1:1.000, 1:10.000 etc.) with an extinction above 0.3 in ELISA. The line at 10.000 marks the titre thought to be acceptable. As a control, saline solution is substituted for an adjuvant in immunisation. The arrows show the times of immunisation. The titre developments of three hens are shown separately.

### Maturation of avidity

Figure 2 shows that FCA is the water-in-oil formulation which induces by far the greatest maturation of avidity. Otherwise only Specol has an effect notably better than the control. The effects of the oil-in-water formulations and the control's are similar.

## Troubleshooting

As mentioned above the success or failure of an immunisation strongly depends on the amount of antigen/adjuvant used. Thus, the chosen amount of PCLS in Subprotocol 1 perhaps was too low to induce an Ab-development. In other investigations (Mittermeier 1995, Erhard et al. 1997, see Table 6), using PCSL (0.25 ml) and human IgG (0.05 mg), antibody titres could be reached comparable to those in Figure 1. The difference between the two studies lies in the amounts of adjuvant (0.1 ml vs. 0.1 - 0.25 ml) and antigen (0.5 - 1.0 mg vs. 0.5 mg - 0.5 mg) used as well as in the immunisation scheme (0/30/60 d).

At this point, attention should be drawn to the special characteristics of antibody titre development in the case of hens. Depending apparently on the immunogenic potency of the antigen, the development may proceed in undulant phases. In other words, after each booster injection the titre rises but then falls again after a while. Nonetheless some developments remain stable for a lengthy period (see Chapter 8).

**Table 6.** Comparison Between the Efficacy of FIA and PCSL in Immunisation of Chickens, Immunisation Schedule

| Adjuvant | FIA | PCSL |
|---|---|---|
| Volume | 0.1 ml/0.25 ml | 0.1 mg/0.25 mg |
| Antigen | human IgG | human IgG |
| Primary | 0.5/5/50 μg/0.5 mg | 0.5/5/50 μg/0.5 mg |
| Boost | 0.5/5/50 μg/0.5 mg | 0.5/5/50 μg/0.5 mg |

Route of application: s.c. and i.m., immunisation interval: 0, 30, 60 days

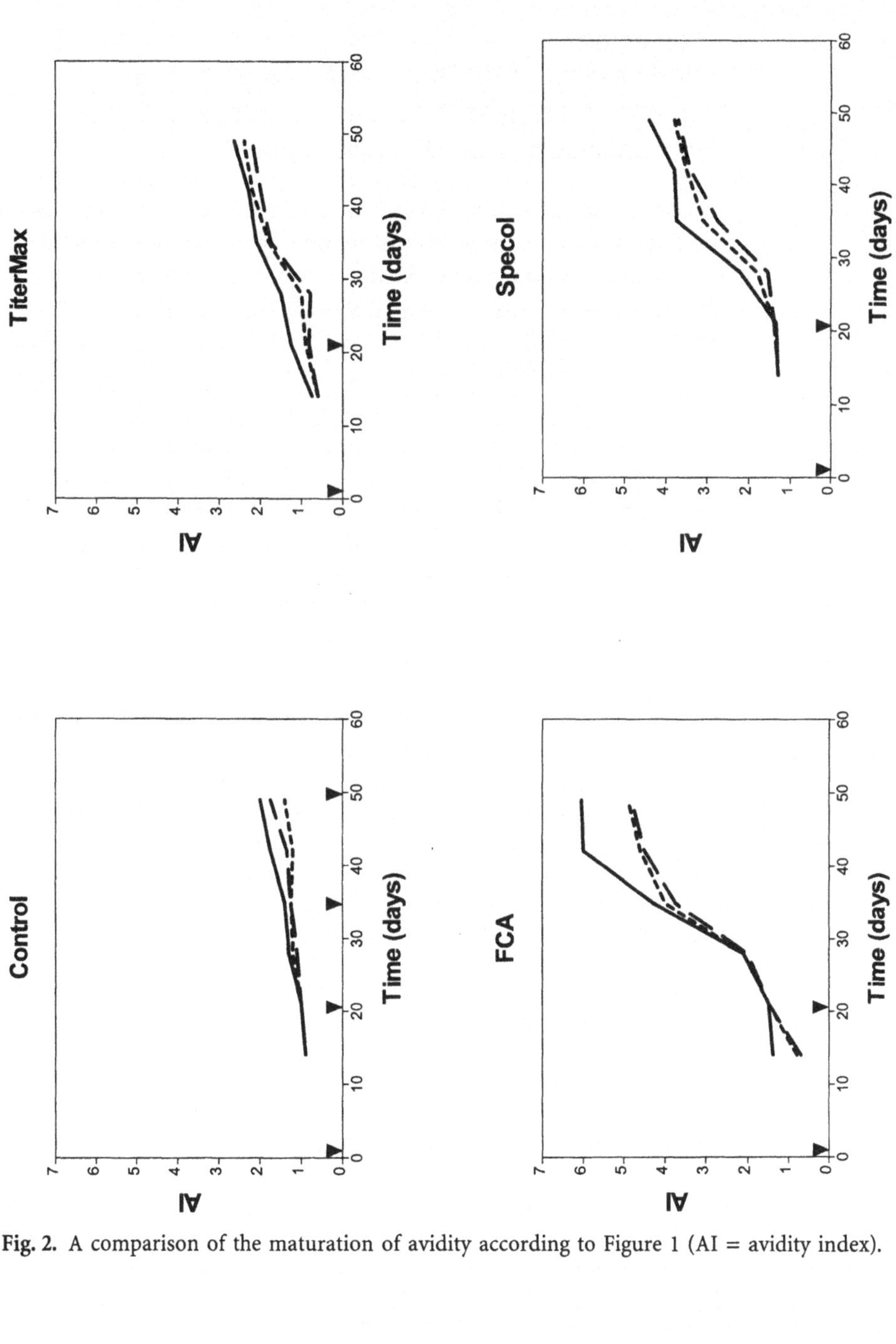


**Fig. 2.** A comparison of the maturation of avidity according to Figure 1 (AI = avidity index).

## Subprotocol 2
## Development of Ab-Titre and Avidity Following an Immunisation With Various Immunogenic Antigens in Combination With Different Adjuvants

In combination with potent immunogens, Specol and TiterMax have proven to be effective adjuvants as compared to FCA, so their suitability for use with corpuscular or weak immunogens should be examined. For this purpose, three antigens with various immunogenicities were applied subcutaneously in combination with each of the named adjuvants. Mammalian IgG (MW: 165 kDa) as a strong immunogen, *Acholeplasma laidlawii* as a corpuscular antigen, and myelin basic protein - MBP (MW: 18 kDa) - as a weak immunogen represented the antigens. In the control groups, a physiological saline solution was applied together with the respective antigen. For details see Subprotocol 1.

### Outline

Table 7 shows the adjuvants used, the concentrations of antigens and the times of immunisation.

**Table 7.** Amount of Antigen and Adjuvant Used for Immunisation

| Adjuvant | FCA | Specol | TiterMax | NaCl |
|---|---|---|---|---|
| Volume | 0.5 ml | 0.5 ml | 0.1 ml | 0.5 ml |
| Antigen | human IgG (Protein, MG: 165.0 kDa) | | | |
| primary immunisation | 1.0 mg | 1.0 mg | 1.0 mg | 1.0 mg |
| boost (21. Day) | 0.5 mg | 0.5 mg | 0.5 mg | 0.5 mg |
| Antigen | MBP (Protein, MG: 18.4 kDa) | | | |
| primary immunisation | 1.0 mg | 1.0 mg | 1.0 mg | 1.0 mg |
| boost (21. Day) | 0.5 mg | 0.5 mg | 0.5 mg | 0.5 mg |
| Antigen | *Acholeplasma laidlawii* (corpuscular antigen) | | | |
| primary immunisation | 1.0 mg | 1.0 mg | 1.0 mg | 1.0 mg |
| boost (21. Day) | 0.5 mg | 0.5 mg | 0.5 mg | 0.5 mg |
| Route of application: s.c. | | | | |

## Materials

As with Subprotocol 1.

## Procedure

As with Subprotocol 1.

## Results

### Titre development (IgG)

Figure 3a shows the Ab-titre development dependent on the respective IgG/adjuvant combination.

### Titre development (Acholeplasma laidlawii)

Figure 3b shows the Ab-titre development dependent on the respective *Acholeplasma laidlawii* /adjuvant combination.

The effects of Specol and FCA after immunisation against mammalian IgG are similar. This good effect of Specol is repeated in the immune response against A. laidawii. Antibodies against the weakly immunogenic MBP with a low molecular weight are produced only after combination of the antigen with FCA or Specol (data not shown) but the titre development is only moderate. A strong response was shown in the case of one hen which also showed strong side effects including inhibited laying performance induced by FCA (data not shown). Under the chosen conditions TiterMax had a notable effect only for IgG.

### Maturation of avidity

Figure 4 shows the influence of adjuvant/antigen combination on avidity 50 days after primary immunisation. The adjuvants Specol and TiterMax induce, like FCA, a good maturation of avidity.

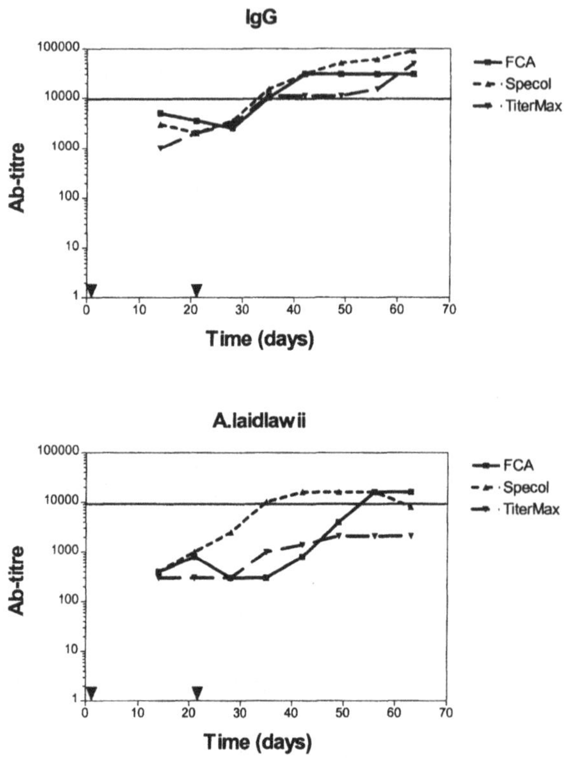


**Fig. 3.** A comparison of the efficacy of various adjuvants on antibody titre-development after immunisation with various immunogenic antigens (a: IgG - soluble antigen, highly immunogen, b: *A. laidlawii* - corpuscular antigen). Shown are the values recorded for each of two hens. Otherwise as in Figure 1.

## Troubleshooting

As with Subprotocol 1.

## Subprotocol 3 The Immune Response of Laying-Hens to Various Immunisation Protocols (Antigen, Adjuvant and Mode of Application)

After various studies (Erhard et al. 1991, Erhard et al. 1994, Hofmann et al. 1996, Schwarzkopf and Thiele 1996, Erhard et al. 1997, Schwarzkopf and Thiele 1997), general principles for

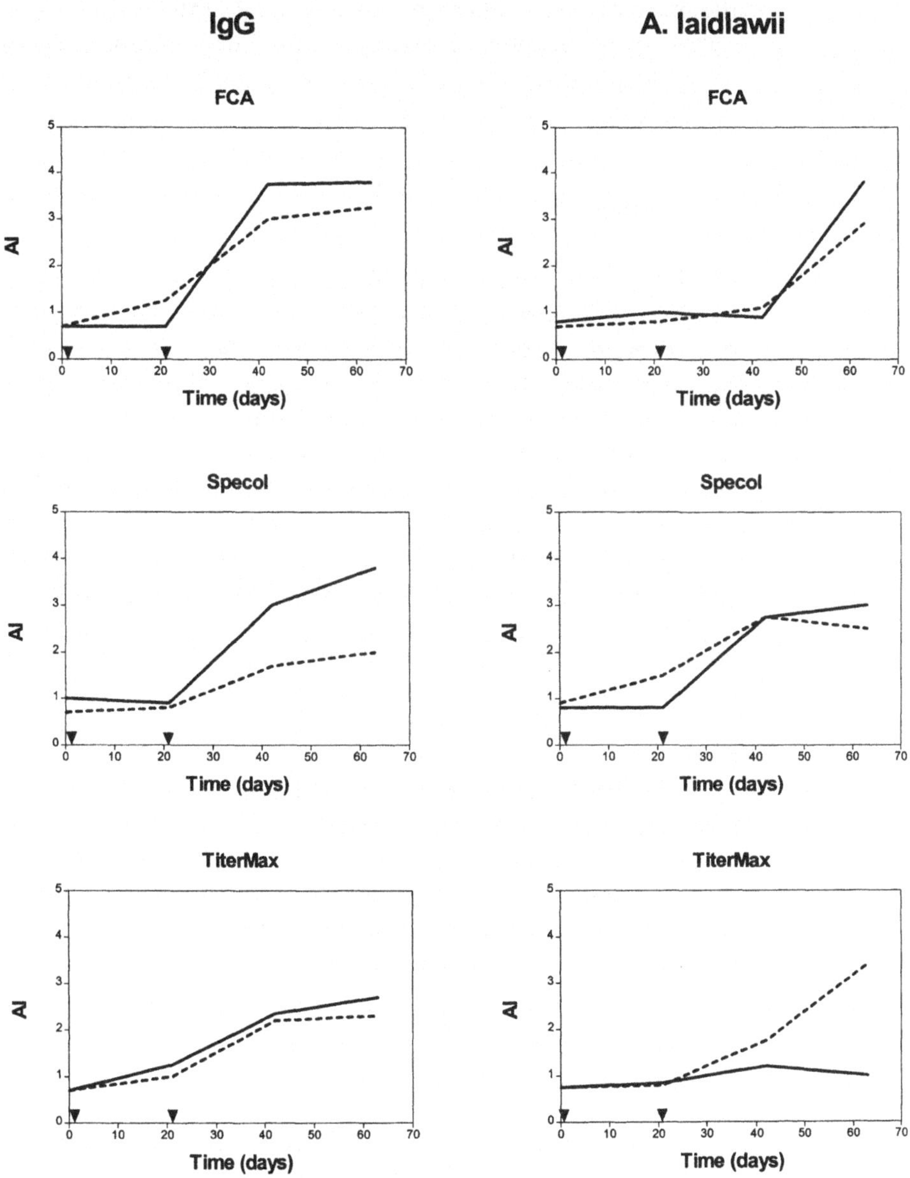


**Fig. 4.** A comparison of the maturity of antibody avidity corresponding with Figure 3. Shown here is the maturation for each of two hens. Otherwise as in Figure 1 or 3.

the immunisation of laying-hens was sought (Mahn 1998). This Subprotocol describes an extensive experiment in order to clarify some questions connected with an optimal immunisation of laying hens (see Erhard et al. 2000).

## Outline

In a large-scale trial, a total of 972 laying-hens was divided into 162 groups of 6 hens and were immunised according to Table 8 once every 4 weeks, over 3 months. The injections were either subcutaneous (in the neck) or intramuscular (in the M. pectoralis). The antigens used were fimbrial antigen colonisation factor 88 from Escherichia coli (K88), recombinant bovine somatotropin (rbST), human immunoglobulin G (hIgG) and low density lipoprotein (LDL) in 4 different concentrations in each case (1 μg, 10μg, 100μg, 1 mg). The antigen controls were given only the respective antigen without any adjuvant, whereas hens injected with only physiological saline acted as the negative controls. The adjuvants used were PCSL or FIA in one of two different concentrations in each case. All these solutions were injected both subcutaneously and intramuscularly in a volume of 0,5 ml.

From the 5th week after immunisation, 1 egg per week was taken from each hen over a period of 11 weeks, yielding over 10,000 samples. The concentration of specific antibodies in the yolks obtained was determined by direct ELISA (see Chapter 5) in comparison with defined standards. The defined ELISA-unit was obtained as follows:

- The total IgY was prepared from a high-titre yolk-pool (see Chapter 4) and the IgY content was measured by means of a sandwich ELISA (see Chapter 5).
- The specific IgY Ab was prepared from the IgY fraction as above by means of affinity chromatography (see Chapter 5) and the IgY concentration was measured by a sandwich ELISA as above.
- The sample of specific IgY (with known concentration) was used to define the ELISA-units by means of a direct ELISA (immobilised antigen + specific Ab, see Chapter 5).

**Table 8.** Amount of Antigen and Adjuvant Used for Immunisation

| Adjuvant | FIA | PCSL |
|---|---|---|
| Volume/amount | 0,1 ml/0.25 ml | 0,1 ml/0.25 mg |
| Antigen | hIgG (Mg: 158 kDa) | |
| Primary | 0,5/5/50 μg/0.5 mg | 0,5/5/50 μg/0.5 mg |
| Boost (30. Day) | 0,5/5/50 μg/0.5 mg | 0,5/5/50 μg/0.5 mg |
| Antigen | K 88 (Mg: 22 kDa) | |
| Primary | 1/10/100 μg/1 mg | 1/10/100 μg/1 mg |
| Boost (30. Day) | 1/10/100 μg/1 mg | 1/10/100 μg/1 mg |
| Antigen | rbSt (Mg: 22 kDa) | |
| Primary | 1/10/100 μg/1 mg | 1/10/100 μg/1 mg |
| Boost (30. Day) | 1/10/100 μg/1 mg | 1/10/100 μg/1 mg |
| Antigen | LDL (Mg: ›1000 kDa) | |
| Primary | 1/10/100 μg/1 mg | 1/10/100 μg/1 mg |
| Boost (30. Day) | 1/10/100 μg/1 mg | 1/10/100 μg/1 mg |

Route of application: s.c. and i.m., immunisation interval: 0,30,60 d (volume of adjuvant/antigen solution: 0.5 ml).

K88 - fimbrial antigen colonisation factor 88 of Escherichia coli

rbSt - recombinant bovine Somatotropin

LDL - low density lipoprotein

- In this way the specific IgY content of an unknown sample can be determined using the same direct ELISA and comparing the extinction obtained with the values defined for the standards.

## Materials

**Equipment**

As with Subprotocol 1.
Additionally:

- Chromatographic columns
- UV-monitor (280 nm)

- Recorder
- Peristaltic pump

**Reagents, solutions**

- CNBr-activated Sepharose
- Respective Antigens

## Procedure

1. Immunisation of laying hens according to Table 8.
2. Preparation of IgY-samples by the water dilution method (see Chapter 4).
3. Preparation of specific Ab by affinity chromatography (see Chapter 4).
4. Determination of the IgY concentration by ELISA (total IgY by sandwich ELISA, specific Ab by direct ELISA, see Chapter 5).

## Results

PCSL and FIA were about equally efficient. In some groups, significantly higher titres could be achieved with PCSL than with FIA (see Figure 5). The effect of both adjuvants also depended on the antigens. Low molecular antigens like rbST and K88 seemed to be less immunogenic than the two higher molecular antigens. An immune response to K88 could not be raised significantly by means of adjuvants. The first booster immunisation led in nearly all groups to a further significant rise in the titres above the levels after the first immunisation, but the rise was nonetheless limited. Especially after immunisation with PCSL, there were further rises in titre.

The influence of the site of application on the immune response was seen after intra-muscular injections of FIA. For PCSL several advantages at the site of subcutaneous injections could be observed.

Save in a few cases, FIA was more effective in doses of 0.1 ml. The optimal dosage of PCSL was found to be at a concentration of 0.25 mg/injection. The highest titre was reached with a dose of

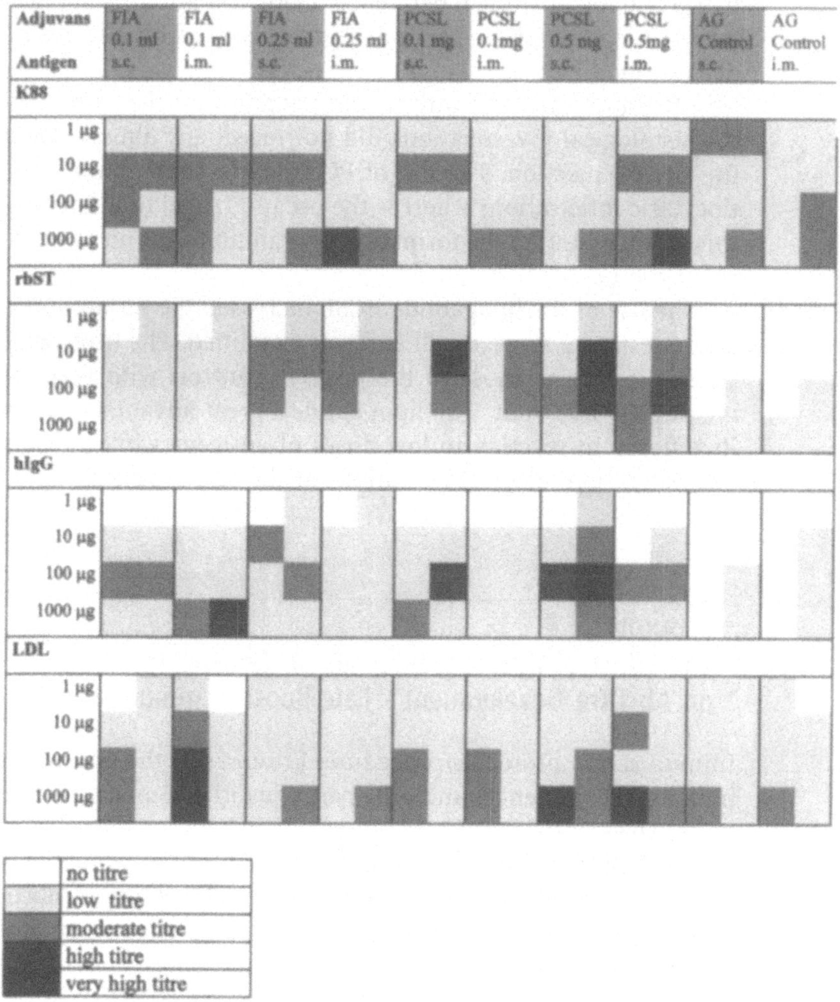


**Fig. 5.** An overview of the distribution of the 972 laying hens into 162 experimental groups (n = 6 hens per group) for the development of an immunisation scheme.
The antigens used were fimbria-antigen colonisation factor 88 from Escheria coli (K88), recombinant bovine somatropin (rbST), human immunoglobulin G (hIgG) and low density lipoprotein (LDL) in various doses. At the same time various doses of the adjuvants $Pam_3Cys$-Ser-$(Lys)_4$(PCSL) and Freunds incomplete adjuvant (FIA) were given. Further groups were given the respective antigen without any adjuvant (AG-control). The injections were subcutaneous (s.c.) or intramuscular (i.m.). the first grade of grey (x/-) gives the value after the first boost, and the second grade (-/x) the value after the second. For the sake of the overview, the values compared are those two weeks after the respective boost. The NaCl control-group showed no specific antibody-titre (data not shown). The levels of titration cannot be given in absolute values, since the extinctions between the various test-systems vary too much to be compared with each other. (A very high titre in one test-system can be equivalent to a moderate titre in another, and vice versa.)

0.01 to 1.0 mg of antigen per hen. Nonetheless after the application of 1 mg there were partly acceptable concentrations of specific antibodies.

After application of antigen without adjuvant the pathological-histological investigations did not reveal any alterations at the site of injection. The use of PCSL led to some lymphohistiocytaric infaltrations whereas the use of FIA led to inflammatory changes and to the formation of granulomeas and to fibrosis of the tissues.

The use of the lipopeptide PCSL increased the humoral immune response to nearly all antigens examined. The titres were similar in size to those of the positive controls with Freunds incomplete adjuvant. The lipopeptide's great advantage lay in its efficacy in reacting to low doses of antigens while causing only slight pathological alterations at the site of injection.

## Subprotocol 4
## The Effect of Different Intervals for Booster Injections on Ab-Titre Development - Late-Booster Immunisation

Immunisation protocols/procedures are as a rule the product of individual experiences and only partly justifiable on a scientific basis. An essential point here is the interval between the primary immunisation and the following booster injections, which may have a significant effect on the immunological memory. Ambrosius and Rudolf (1990) mention that the interval between first and second immunisation is generally too short and consequently, it would be advisable to extend this interval.

### Outline

The following table shows an overview of the adjuvants used, the antigen concentrations used and the time of immunisation.

**Table 9.** Immunisation Schedule

| | |
|---|---|
| Antigen: human IgG (protein, MW: 165 kDa) | |
| Primary Immunisation: 1.0 mg, Boost 0.5 mg | |
| FCA | Day: 0 / 21 / 109 / 133 |
| Specol | Day: 0 / 21 / 109 / 133 |
| TiterMax | Day: 0 / 21 / 109 / 133 |
| PCSL | Day: 0 / 21 / 109 / 133 |
| NaCl | Day: 0 / 21 / 109 / 133 |

## Materials

As with Subprotocol 1.

## Procedure

As with Subprotocol 1.

## Results

Investigations into the course of Ab-titre development after 200 days show clearly that the period in which a titre reaches a satisfying level against strongly immunogenic protein is increased from about 50 to 150 days, if the conventional first immunisation phase is followed by a later booster immunisation using the same adjuvant. This applies to FCA to the same extent as to Specol (Figure 6). Comparable observations have been made by Hlinak et al. (1996).

## Troubleshooting

As with Subprotocol 1.

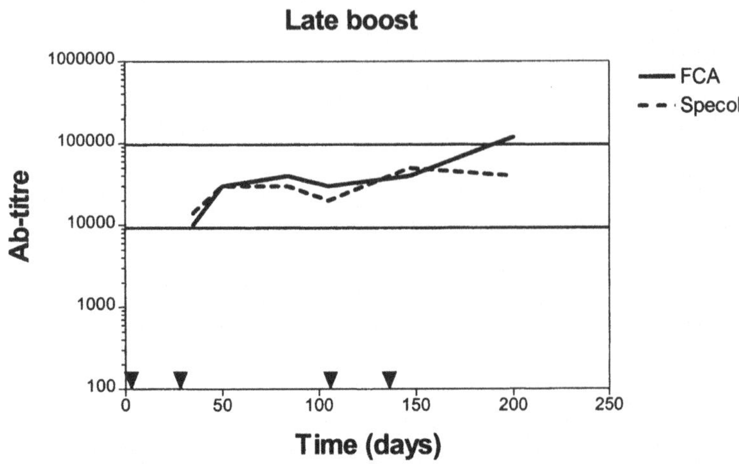


**Fig. 6.** Influence of the interval between boosts on the development of the antibody titre after immunisation with human IgG (n = 2). Between the 1st and the 2nd boosts lay an interval of more than 2 months (see the arrow). The later boost produced a clear rise in the antibody titre, especially in the case of FCA.

## Subprotocol 5
## The Effect on Ab-Titre Development After Consecutive Immunisation With Different Adjuvants

There are indications of a positive influence on Ab-titre development by the consecutive use of two adjuvants (Schwarzkopf and Thiele 1996). In this protocol some data are shown resulting from the combined use of different adjuvants.

### Outline

The following table gives an overview of the adjuvants used, the antigen concentration and the time of immunisation.

### Materials

As with Subprotocol 1.

**Table 10.** Consecutive Use of Different Adjuvants

Antigen: IgG (protein, MW: 165 kDa)

Primary Immunisation: 1.0 mg

Booster Immunisation: 0.5 mg

Volume per Immunisation: FCA: 0.5 ml; Gerbu 0.1 ml; NaCl: 0.5 ml

| | | | |
|---|---|---|---|
| FCA | Day: 0 / 21* | FCA | day: 100 / 121 |
| FCA | Day: 0 / 21 | Gerbu | day: 100 / 121 |
| Gerbu | Day: 0 / 21 | Gerbu | day: 100 / 121 |
| Gerbu | Day: 0 / 21 | FCA | day: 100 / 121 |
| NaCl | Day: 0 / 21 / 35 / 49 | NaCl | day: 100 / 121 |

* "conventional immunisation"

## Procedure

As with Subprotocol 1.

## Results

After a conventional immunisation with FCA (day 0 + 21), a plateau with useful titres is formed over a period of at least 50 days. A later booster immunisation on days 100 and 121 leads to a sudden fourfold rise in titre if FCA or the Gerbu adjuvant is used. This plateau is higher and more stable than the first and decreases slowly over 200 days (see Figure 7).

A conventional immunisation alone with the Gerbu adjuvant did not produce useful titres. A conventional primary immunisation with FCA, combined with a later booster injection with the Gerbu adjuvant, leads to a greater increase in titre than does a late boost with FCA.

## Troubleshooting

As outlined above, the crucial point of an immunisation ist the interaction of three variables (antigen and adjuvant, both varying in nature and concentration; animal, varying species). Thus,

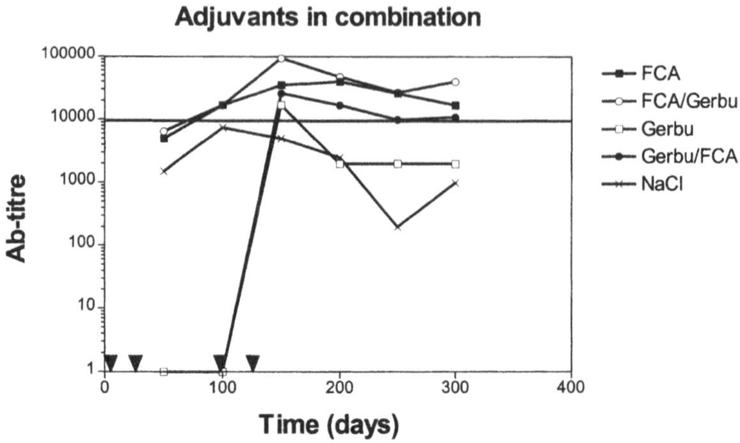


**Fig. 7.** Development of the antibody titre after immunisation with human IgG together with various adjuvants. For the 2nd and 3rd boost (on day 100 or 121) a different adjuvant was used in each case (see also Table 10).

the results presented can only be generalized within certain limits.

## Comments

### Adjuvants compared in terms of their side effects

Tables 11 and 12 show tissue changes after the application of various adjuvants. The sizes of abscesses do correspond with the applied volume of oil-containing adjuvants and are about 10-20 times the size of the injected volume. Abscesses are formed after subcutaneous and intramuscular injections.

The pathological-histological investigation revealed nothing in the animals immunised without any adjuvant; the use of PCSL led to low-grade and, occasionally, to medium-grade infiltrates with lymphocytes and histiocytes, showing an immune reaction weaker after subcutaneous injections; and the use of FIA led plainly to inflammatory changes and sometimes to the formation of granulomas and to fibrosis.

Subcutaneously localised alterations can be shifted freely and affect the tested animal less than do those after intramuscular injections. In general all oil-containing adjuvants tend to induce abscesses, though the pathological changes after repeated use of the adjuvant Specol, containing mineral oil, are relatively mild. The adjuvants with least notable side-effects are Gerbu and the lipopeptide PCSL (microscopic and macroscopic determination according to Schwarzkopf and Thiele 1996, see also Schmidt et al. 1996).

The PCSL data are not shown in Subprotocol 1 because there is no difference to the respective control.

Nearly all adjuvants are more tissue tolerable than is FCA, the adjuvant with most undesirable side-effects.

**Table 11.** Pathological Changes After Application (s.c.) of Antigen and Adjuvant According to Subprotocol 1

| | Immunisation | Booster immunisation |
|---|---|---|
| I. Water-in-oil formulations | | |
| FCA | +++ | +++ |
| Specol | + | + |
| TiterMax | + | + |
| II. Oil-in-Water formulations | | |
| ABM-System | + | + |
| Neoparasec® | + | + |
| Gerbu adjuvant | – | – |
| Lipopeptide (PCSL) | – | – |

– no macroscopic changes; + mild, +++ severe changes

**Table 12.** Pathological Changes After Application (s.c) of Antigen and Adjuvant According to Subprotocol 2

| | IgG | | MBP | | A. laidlawii | |
|---|---|---|---|---|---|---|
| | Primary | Booster inj. | Primary | Booster inj. | Primary | Booster inj. |
| FCA | +++ | +++ | ++ | ++ | ++ | ++ |
| PCSL | – | – | +++ | +++ | – | – |
| Specol | + | + | + | + | + | + |
| TiterMax | + | + | + | + | + | + |
| NaCl | – | – | – | – | – | – |

– no macroscopic changes; + mild, ++ moderate, +++ severe changes

Moreover, at least in the case of lipopeptide PCSL, there seem to be side-effects depending on antigens. Whereas PCSL, combined with IgG or *A. laidlawii*, induces no essential tissue changes, abscesses arise with the combination PCSL/MBP. The reason is open to speculation. Myelin basic protein for instance, which is a known target antigen for the T- cellular immune reaction in experimental allergic myelitis (animal model of multiple sclerosis), may lead to other effector mechanisms of the immune response with consequent harm to tissues when combined with PCSL.

## Comparison of various adjuvants in regard to labour input/benefit relation

The choice of a suitable adjuvant depends not only on its efficacy and side-effects but also on its price and the difficulties in preparing it. The following table gives an overview of the costs / ease of treatment / efficacy / side-effects of various adjuvants used on hens.

## Conclusions for a successful scheme for immunisation of hens

Using suitable adjuvants and adapted immunisation schemes, hens react with the production of antibodies which in titre, avidity and binding specificity do not differ from mammalian ones, yet the yield is much greater.

As explained at the beginning, only general recommendations can be given as regards the choice of an optimal adjuvant:

**Table 13.** Comparison of the Cost/Benefit Relations and Side Effects of Various Adjuvants

| | FCA | Specol | TM | ABM | Gerbu | PCSL |
|---|---|---|---|---|---|---|
| high costs | | | X | X | X | X |
| easy handling | | X | | X | X | X |
| good immune response | X | X | X | | | X |
| more severe side effects | X | | | | | |

Generally, using the adjuvants FCA and Specol, both containing mineral oils, the highest titres and strongest avidities are reached. But in the case of FCA these are accompanied by undesirable side-effects like abscesses and degenerative changes in muscle tissue. Specol, like FCA, is an adjuvant composed of non-metabolisable mineral oils, but there are not so strong side-effects as with FCA. According to present experience, only FCA induces a useful but variable immune response to antigens with low molecular weights not bound to carriers. In the case of corpuscular and high molecular antigens, Specol is a good alternative to FCA for the immunisation of laying hens, as it has only mild side-effects and induces an immune response which is long-lasting and comparable to FCA's in strength.

In view of its side-effects, FCA should be used only with weak immunogens. It should be considered whether the side-effects can be lessened by replacing FCA with FIA or by choosing a ratio lower than 50/50 (v/v) for antigen/adjuvant. See also the recommendations of the ECVAM-reports on this theme and guidelines of the Swiss Bundesamt für Veterinärwesen (Schade et al. 1996, Leenaars et al. 1999, Anonymous 1999).

Commercially available alternative adjuvant are Specol, PCSL and TiterMax. Specol seems to have an immunogenic effect similar to FCA's. In view of present results, the successful use of PCSL seems to depend on keeping its concentration within certain limits (250 μg/injection, s.c.). Adjuvants with a small part of non-metabolisable mineral oils or with metabolisable ones like TiterMax, the ABM system, Neoparasec® and the Gerbu adjuvant did not have notable side-effects under the chosen experimental conditions. Used alone, they were not effective enough.

#### Standard protocol

Various factors affect the immunisation of hens (see Schade et al., 1992). The primary immunisation may be followed by another 3-4 weeks later, or several immunisations may follow one another at short intervals, then be followed by the final booster injection after 1-2 months. Late boosts (3-4 months) may lead to a drastic rise in the antibody titre.

**Table 14.** Recommendations for Immunisation of Chickens

| | | | |
|---|---|---|---|
| Antigen | | 0.10 - 1.00 mg | Sometimes 10 µg sufficent |
| Adjuvant | FIA | 0.10 - 0.25 ml | Usually i.m. or s.c. |
| | Specol | 0.50 ml | s.c. |
| | PCLS | 250 µg | Preferably subcutaneous |
| | FCA | 0.10 - 0.25 (0.5) ml | Only with weakly immunogenic antigen |
| Injections: | | | |
| - volume | | 0.5 ml (-1 ml) | |
| - intervals | | 4-8 weeks | |
| - number | | 2 (or more if necessary) | |
| - route | | i.m. or s.c. | |

As a rule the antigen-emulsion is injected subcutaneously or into the breast-muscle (FCA/FIA).

Experiments suggest that an antigen concentration of 0.1 - 1 mg is favourable for the immunisation of hens. In particular cases, a lower or higher concentration may be more suitable. In the case of immunisation with peptide-carrier conjugates, it may be advantageous to begin with higher concentrations (0.25 mg peptide/immunisation).

Supported by these and other results and a concern for animal protection, discussion within an international ECVAM-workshop led to a general immunisation scheme for laying-hens (Schade et al. 1996). Table 14 shows the recommended protocol.

## References

Allison AC, Byars NE, Waters RV (1986) Immunologic adjuvants: efficacy and safety considerations. In: Nervig RM, Gough PM, Kaeberle ML, Whetstone CA (eds) Advances in carriers and adjuvants for veterinary biologics. Iowa State University Press, Ames, Iowa, USA, pp 91-103

Alving CR (1997) Liposomes as adjuvants for vaccines. In: Levine MM, Woodrow GC, Kaper JB, Cobon GS (eds) New generation vaccines, 2nd edn, Marcel Decker, New York, pp. 207-213

Ambrosius R, Rudolph W (1990) Grundriß der Immunbiologie, Second Edition, Gustav Fischer Verlag, Stuttgart New York, pp. 29-31

Amyx HL (1987) Control of animal pain and distress in antibody production and infectious disease studies. J. Am. Vet. Med. Assoc. 191:1287-1289
Anonymous (1999) Fachgerechte und tierschutzkonforme Antikörperproduktion in Kaninchen, Hühnern und Labornagetieren (Richtlinie Tierschutz 3.04). Bundesamt für Veterinärwesen, 3097 Liebefeld-Bern, 8.März 1999, ho-800.116-3.04
Azuma I, Sugimura K, Taniyama T, Yamawake M, Yamamura Y, Kusumoto S, Okada S, Shiba T (1976) Adjuvant activity of mycobacterial fractions: adjuvant activity of synthetic N-acetyl muramyl-dipeptide and the related compounds. Infect. Immun. 14:18-27
Beebe GW, Simon AH, Vivona S (1972) Long-term mortality follow-up of army recruits who received adjuvant influenza virus vaccine in 1951-1953. Amer. J. Epidemiol. 95:337-344
Behn I, Hommel U, Oertel M, Hauschild S (1996) Kinetics of IgY formation after immunisation of hens with different protein antigens. ALTEX Suppl. 13:18-21
Bennett B, Check IJ, Olson MR, Hunter RL (1992) A comparison of commercially available adjuvants for use in research. J. Immunol. Methods 153:31-40
Boersma WJA, Bogaerts WJC, Bianchi ATJ, Claassen E (1992) Adjuvant properties of stable water-in-oil emulsions: evaluation of the experience with Specol. Research in Immunology 143:503-512
Boersma WJA, Haaijman JJ, Claassen E (1993) Use of synthetic peptide determinants for the production of antibodies. In: Cuello AC (ed) Immunohistochemistry II, IBRO, John Wiley & Sons, Chichester New York Brisbane Toronto Singapore, pp. 1-77.
Bokhout BA, Van Gaalen C,Van der Heijden PJ (1981) A selected water-in-oil emulsion: composition and usefulness as an immunological adjuvant. Vet. Immunol. Immunopath., 2:491-500
Bomford R (1985) Adjuvants. Animal Cell Biotechnology 2:235-250
Broderson JR (1989) A retrospective review of lesions associated with the use of Freund's Adjuvant. Lab. Anim. Sci. 39:400-405
Dunnigton EA, GROSS WB, Larsen CT, Siegel PB (1992) Antibody response to combination of antigens in populations of chickens from diverse genetic backgrounds. Arch. Geflügelk. 57:171-174
Ellouz F, Adam A, Ciorbaru R, Lederer E (1974) Minimal structural requirements for adjuvant activity of bacterial peptidoglycan derivates. Biochem. Biophys. Res. Commun. 59:1317-1325
Erhard MH, Kellner J, KühlmannR, Lösch U (1991) Influence of various adjuvants on the synthesis of specific antibodies of chicken, sheep and rabbit following immunization with an hapten. J. Vet. Med. A 38:21-27
Erhard MH, Kellner J, Bessler WG Lösch U (1994) Ein Lipopeptid als Adjuvans zur Immunisierung von Hühnern. Biochemica-Information 94:13-14
Erhard MH, Schmidt P, Hofmann A, Stangassinger M, Lösch U (1996) Production and purification of mouse subclass specific chicken egg yolk antibodies using a new indirect affinity chromatography method with protein G. ALTEX Suppl. 13:66-69

Erhard MH, Hofmann A, Bergmann J, Mittermeier P, Kaufmann P, Wiesmüller K-H, Bessler WG, Lösch U (1997) The lipopeptide $Pam_3$Cys-Ser-$(Lys)_4$: an alternative adjuvant to Freund's adjuvant for the immunisation of chicken to produce egg yolk antibodies. ATLA 25:173-181

Erhard MH, Mahn K, Schmidt P, Oltmer S, Preisinger R, Zinsmeister P, Stangassinger M (2000) Evaluation of various immunisation procedures in laying hens induce high amounts of specific egg yolk antibodies. ATLA 28:63-80

Finger H (1964) Das Freundsche Adjuvans. Wesen und Bedeutung. In: Heymann G (ed). Arbeiten aus dem Paul-Ehrlich-Institut, dem Georg-Speyer-Haus und dem Ferdinand-Blum-Institut zu Frankfurt a. M.. Gustav Fischer Verlag, Stuttgart. Heft 60

Freund J (1951) The effect of paraffin oil and mycobacteria on antibody formation and sensitization. Am. J. Clin. Pathol. 21:645-656

Freund J, Casals J, Hosmer EP (1937) Sensitization and antibody formation after injection of tubercle bacilli and paraffin oil. Proc. Soc. Exp. Biol. Med. 37:509-513

Freund J, McDermott K (1942) Sensitization to horse serum by means of adjuvants. Proc. Soc. Exp. Biol. Med. 49:548-553

Freund J, Bonanto MV (1944) Effect of paraffin oil, lanolin-like substances and killed tubercle bacilli on immunization with diphtheric toxoid and Bact. typhosum. J. Immunol. 48:325-334

Glenny AT, Buttle GA, Stevens MF (1931) Rate of disappearance of diphtheria toxoid injected in rabbits and guinea pigs: toxoid precipitates with alum. J. path. Bact. 34:267-275

Glenny AT, Pope GA, Waddington H, Wallace W (1926) Immunological notes XXIII. The antigenic value of toxoid precipitated by potassium alum. J. Path. Bact. 29:38-39

Herbert WJ (1974) Veterinary Immunology. Blackwell Scientific Publications, London. pp. 205-206

Hlinak A, Schrödl W, Witt S, Schade R, Krüger M (1996) Production of egg yolk antibodies against human cell-associated antigens. Altex, Suppl. 13:76-79

Hillemann MR (1967) Considerations for safety and application of emulsified oil adjuvants to viral vaccines. In: Regmay RH, Hennessen W, Ikic D, Unger J (eds) Symposia Series in Immunobiological Standardization, Vol.6, International Symposium on Adjuvants of Immunity. S. Karger, New York, USA. 18

Hofmann A., Erhard MH, Schmidt P, Bessler W, Wiesmüller K-H, Zinsmeister P, Stangassinger M, Lösch U (1996) Lipopeptide als Adjuvantien bei der Immunisierung von Legehennen. ALTEX Suppl.13:26-29

Hudson L, Hay FC (1991) Practical Immunology. Third edition. Blackwell Scientific Publications, Oxford, London, Edinburgh, Boston, Melbourne, Berlin, Paris, Vienna

Jilg W (1996) Schutzimpfungen. In: Baenkler HW (ed) Medizinische Immunologie. Ecomed, Landsberg.

Johnson AG, Gaines S, Landy M (1956) Studies on the O antigen of Salmonella typhosa. V. Enhancement of the antibody response to protein antigens by the purified lipopolysaccharide. J. Exp. Med. 103:225-246

Kaeberle ML (1986) Function of carriers and adjuvants in induction of immune responses. In: Nervig RM, Gough PM, Kaeberle ML, Whetstone CA (eds). Advances in carriers and adjuvants for veterinary biologics. Iowa State University Press, Ames, Iowa, USA. 11-23

Kellner J (1990) Einfluß verschiedener Adjuvantien auf die Antikörpersynthese nach der Immunisierung mit einem Hapten am Beispiel der Spezies Schaf, Kaninchen, Huhn und Maus. Vet. med. Diss., Universität München

Kellner J, Erhard MH, Schranner I, Lösch U (1992) The influence of various adjuvants on antibody synthesis following immunization with an hapten. Biol. Chem. Hoppe-Seyler 373:51-55

Landsteiner K (1945) The specificity of serological reactions. Harvard University Press, Cambridge Mass. USA

Leenars PPAM, Hendriksen CFM, Angulo AF, Koedam MA, Claassen E (1994) Evaluation of several adjuvants as alternatives to the use of Freund's adjuvant in rabbits. Vet. Immunol. Immunopath. 40:225-241

Leenars PPAM, Hendriksen CFM, Koedam MA, Claassen I, Claassen E (1995) Comparison of adjuvants for immune potentiating properties and side effects in mice. Vet. Immunol. Immunopath. 48:123-138

Leenars PPAM, Hendriksen CFM, de Leeuw WA, Carat F, Delahaut P, Fischer R, Halder M, Hanly WC, Hartinger J, Hau J, Lindblad EB, Nicklas W, Outschoorn IM, Stewart-Tull DES (1999) The production of polyclonal antibodies in laboratory animals. The report and recommendations of ECVAM workshop 35. ATLA 27:79-102

Lösch U, Schranner I, Wanke R, Jürgens L (1986) The chicken egg, an antibody source. J. Vet. Med. B 33:609-619

Lübke A (1990) Änderung des Immunstatus (Serumantikörper) beim Kaninchen nach parenteraler Adjuvans- und Antigengabe. Vet. med. Diss., Freie Universität Berlin

MacDonald RA, Hosking CS, Jones CL (1988) The measurement of relative antibody affinity by ELISA using thiocyanate elution. J. Immunol. Methods 106:191-194

Mahn K (1998) Etablierung eines Immunisierungsschemas für Legehennen. Vet. med. Diss, Universität München

Mallon, F. M., M. E. Graichen, B. R. Conway, M. S. Landi, and H. C. Hughes, 1991. Comparison of antibody response by use of synthetic adjuvant system and Freund complete adjuvant in rabbits. Am. J. Vet. Res. 52, 1503-1506

Mayr A, Eissner G, Mayr-Bibrack B (1984) Handbuch der Schutzimpfungen in der Tiermedizin. Verlag P. Parey, Berlin und Hamburg

Mittenbühler K, Loleit M, Baier W, Fischer B, Sedelmeier E, Jung G, Winkelmann G, Jacobi C, Weckesser J, Erhard MH, HofmannA, Bessler WG, Hofmann P (1997) Drug specific antibodies: T-cell epitope-lipopeptide conjugates are potent adjuvants for small antigens in vivo and in vitro. Int. J. Immunnopharm. 19:277-287

Mittermeier M (1995) Das Lipopeptid $Pam_3$Cys-Ser-$(Lys)_4$ - eine Alternative zu Freundschem komplettem Adjuvans (FCA) bei der Immunisierung von Legehennen zur Gewinnung von Dotterantikörpern. Doktor-Thesis, Tierärztliche Fakultät der Ludwig-Maximilians-Universität München

Morein B, Villacres-Eriksson M, Kerblom L, Lövgren K (1993) Initiation of immune response with ISCOM. In: Gregoriadis G, McCormack B, Allison AC, Poste GC (eds) New Generation Vaccines: The Role of Basic Immunology,. Plenum Press, New York, pp.

Nail SL, White JL, Hem SL (1976) Structure of aluminium hyroxide gel II: aging mechanism. J. Pharm. Sci. 65:1192-1195

Nakashima I, Nagase F, Yokochi T, Ohta M, Kato N (1979) Adjuvant actions of polyclonal lymphocyte activators. I. Comparison and characterization of their actions in antibody response to deaggregated bovine serum albumin. Cell. Immunol. 46:69-76

Paterson PY (1966) Experimental allergic encephalomyelitis and auto-immune disease. Adv. Immunol. 5:131-201

Potter M (1977) Antigen-binding myeloma proteins of mice. Adv. Immunol. 25:141-145

Ramon G (1925) Sur l'augmentation anormale de l'antitoxine chez les chevaux producteurs de sérum antidiphtérique. Bull. Soc. Centr. Med. Vet. 101:227-234

Roitt IM, Brostoff J Male DK (1987) Kurzes Lehrbuch der Immunologie. G. Thieme Verlag, Stuttgart New York

Sanchez Y, Ionescu-Matiu I, Dresman G, Kramp W, Six HR, Hollinger FB, Melnick JL (1980) Humoral and cellular immunity to hepatitis B virus-derived antigens: comparative activity of Freund's complete adjuvant, alum, and liposomes. Infect. Immun. 30:728-733

Sasse M, Krüger M, Schade R, Hlinak A (1998) Herstellung und Charakterisierung aviärer vitelliner Antikörper gegen Lipopolysaccharide und Lipid A. 1. Mitteilung: Induktion und Präparation von spezifischen Dotterantikörpern (IgY) gegen Endotoxine. BMTW 111:121-126

Schade R, Schniering A, Hlinak A (1992) Die Gewinnung spezifischer, polyklonaler Antikörper aus dem Dotter von Hühnereiern als Alternative zur Immunisierung von Säugern - eine Übersicht. ALTEX 9:43-56.

Schade R, Staak C, Hendriksen C, Erhard MH, Hugl H, Koch G, Larsson A, Pollmann W, Van Regenmortel M, Rijke E, Spielmann H, Steinbusch H, Straughan D (1996) The production of avian (egg yolk) antibodies: IgY. ATLA 24:925-934.

Schade R, Henklein P, Hlinak A, de Vente J, Steinbusch H (1996) Specificity of chicken (IgY) versus rabbit (IgG) antibodies raised against cholecystokinin octapeptide (CCK-8). Altex, Suppl. 13:80-85

Schecklies E (1996) Polyklonale Antikörper. Eine Einführung in die Theorie und Praxis der Antikörperherstellung. VCH Verlagsgesellschaft mbH, Weinheim New York Basel Cambridge Tokyo

Schmidt P, Erhard MH, Wanke R, Hofmann A, Schmahl W (1996) Gewebsveränderungen nach Anwendung verschiedener Adjuvantien für die Immunisierung. ALTEX Suppl. 13:30-34

Schütt C, Schilling T, Grünwald U, Schönfeld W, Krüger C (1992) Endotoxin-neutralizing capacity of soluble CD14.Res. Immunol. 143:71-78

Schwarzkopf C (1994) Gewinnung und immunologische Charakterisierung speziesspezifischer IgY sowie deren Einsatz zur Bestimmung der Wirtstierart aus dem Abdominalblut hämatophager Insekten. Vet. med. Diss., Freie Universität Berlin

Schwarzkopf C, Thiele B (1996) Effectivity of alternative adjuvants in comparison to Freund's complete adjuvant, ALTEX, Suppl. 13:22-25

Schwarzkopf C, Thiele B (1997) Chicken IgY: efficacy of alternative adjuvants compared to FCA. In: van Zutphen LFM, Balls M (eds) Animal alternatives, welfare and ethics, Elsevier, Amsterdam Lausanne New York Oxford Shannon Tokyo, pp. 983-988

Snippe H, Kamp E (1975) Cell co-operation and hapten-carrier complexes. Immunology 29:675-685

Sprick-Sanjosé Messing A (1990) Studien zur Adjuvanswirkung beim Haushuhn (Gallus gallus). Vet. med. Diss., Universität München

Steiner JW, Langer B, Schatz DL (1960) The local and systemic effects of Freund's adjuvant and its fractions. Arch. Pathol. 70:424-434

Sugimoto M, Germain RN, Chedid L, Benacerraf B (1978) Enhancement of carrier-synthetic helper T-cell function by the synthetic adjuvant, N-acetyl muramyl-L-alanyl-D-isoglutamine (MDP). J. Immun. 120:980-982

Tal C, Laufer A (1960) Amyloidosis in mice following injections with Freund's adjuvant, its components separately or in homologous liver-adjuvant mixture. Br. J. Exp. Pathol. 41:605-609

Tam JP (1988) Synthetic peptide vaccine design: synthesis and properties of a high-density multiple antigenic peptide system. Proc Nat Acad Sci USA 85:5409-5413

Thein P (1988) Adjuvantien: Immunmodulation und Immunstimulation. Vet. Med. Nachr. 59:3-8

Wanke R, Schmidt P, Erhard MH, Sprick-Sanjose Messing A, Stangassinger M, Schmahl W, Hermanns W (1996) Freundsches komplettes Adjuvans beim Huhn: effiziente Immunstimulation bei gravierender lokaler inflammatorischer Reaktionen. J. Vet. Med. A 43:243-253

Wiesmüller KH, Bessler WG, Jung G (1983) Synthesis of the mitogenic S-[2,3-Bis(palmitoyloxy)propyl]-N-palmitoylpenta-peptide from Escherichia coli lipoprotein. Hoppe-seyler's Z. Physiol. Chem. 364:593-606

World Health Organisation (1976) Immunological adjuvants. Technical Report Series No. 595, Geneva

## Suppliers

**FCA/TiterMax** SIGMA
Tel.: (0049) 89-6513-1554
FAX: (0049) 89-6513-1530
E-mail: DeTSBio@eurnotes.sial.com

**Specol** Instituut voor Dierhouderij en Diergezondheid (ID-DLO)
Tel.: 0320-23 82 38
FAX: 0320-23 80 50
E-Mail: postkamer@id.dlo.nl

**ABM-N, ABM-S** Linaris
Tel.: (0049) 9342-39499
FAX: (0049) 9342-39994

**Gerbu adjuvant** Gerbu Biotechnik GmbH
Tel.: (0049) 6223-9513-0
FAX: (0049) 6223-951319
E-mail: gerbu@t-online.de

**Lipopeptide PCSL** Boehringer Mannheim
Tel.: (0049) 621-759-85 45
FAX: (0049) 621-759-85 09
E-mail: BiochemInfo_de@bmg.boehringer-mannheim.com

**Additionally: Complete adjuvant**
**(based on lipo-polysacharides from blue-green algae)** BioGenes
Tel.: (0049) 30-65 76 23 96
FAX: (0049) 30-65 76 23 97
E-mail: biogenes@berlin.sireco.net

# Isolation of IgY from Yolk

CHRISTIAN STAAK, CHRISTINE SCHWARZKOPF, INGRID BEHN, UNDINE HOMMEL, ANDREAS HLINAK, RÜDIGER SCHADE, and MICHAEL ERHARD

## Introduction

### Constitution of the Yolk

The yolk in hens' eggs consists of dry matter (51,3 %) and water (48,7 %) in nearly equal proportions (Siewert and Bronch, Handbuch der Tierernährung, 1972; Romanoff und Romanoff, The Avian Egg, 1949). The dry matter is made up as shown in Table 1.

Regarding its constitution, yolk can be viewed as being an oil-in-water emulsion with a watery portion containing proteins and a dispersal portion of so-called yolk-granules and lipid-drops (Fischer, 1996; Fischer et al., 1996).

✉ Christian Staak, Bundesinstitut für gesundheitlichen Verbraucherschutz und Veterinärmedizin, Diedersdorfer Weg 1, Berlin, 12277, Germany (*phone* +49-30-8412-2053; *fax* +49-30-8412-2956 ; *e-mail* christian.staak@bgvv.de)
Christine Schwarzkopf, Zentrales Institut des Sanitätsdienstes der Bundeswehr Koblenz, Laborabt. II (Vet.Med), Generaloberst-Beck-Str. 17, Mainz, 55129, Germany
Ingrid Behn, Universität Leipzig, Fakultät für Biowissenschaften, Pharmazie und Psychologie, Talstr. 33, Leipzig, 04103, Germany
Undine Hommel, Universität Leipzig, Fakultät für Biowissenschaften, Pharmazie und Psychologie, Talstr. 33, Leipzig, 04103, Germany
Andreas Hlinak, Universitätsklinikum Charité der Humboldt-Universität, Institut für Pharmakologie und Toxikologie, Dorotheenstr. 94, Berlin, 10117, Germany
Rüdiger Schade, Universitätsklinikum Charité der Humboldt-Universität, Institut für Pharmakologie und Toxikologie, Dorotheenstr. 94, Berlin, 10117, Germany
Michael Erhard, Universität Leipzig, Veterinärmedizinische Fakultät, Veterinär-Physiologisches Institut, An den Tierkliniken 7, Leipzig, 04103, Germany

Table 1. Constitution of Egg Yolk

| | |
|---|---|
| Protein | 16.6% |
| Fats and lipoids | 32.6 % |
| Carbohydrates | 1.0 % |
| Anorganic matter | 1.1 % |

The proteins in the yolk can be divided roughly into four fractions:

1. The vitellin or lipovitellin fraction (47.5 %), a phosphorus-containing nucleo-albumin (400 kDa) with subsidiary units having molecular weights between 35 and 140 kDa (Mehner and Hartfield, 1983).
2. The vitellelin fraction (38.6 %), with less phosphorus and more lipids than the vitellin fraction.
3. The phosvitin fraction (4.3 %) - very rich in phosphorus and without lipids - a protein with a Mw of 36 kDa (Mehner and Hartfield, 1983).
4. The livetin fraction (9.6 %) with soluble proteins like globulin.

Fraction 4 is very heterogeneous, containing about 10 proteins, about which various claims are made (Mehner and Hartfield, 1983). The original differentiation into alpha-, beta- and gamma-livetin (Martin et al., 1959) has been replaced as follows (Table 2):

The concentration and distribution of immunoglobulins in the serum and egg are as follows (Table 3, Rose et al., 1974):

Table 2. Composition of the Egg Yolk Livetin Fraction

| | |
|---|---|
| alpha-livetin | the same as plasma-albumins (54 kDa) (Williams, 1962) |
| beta-livetin | alpha-2-glycoprotein (Hatta et al., 1990) |
| gamma-livetin | the same as serum gamma-globulins, together with transferrin (77 kDa, Williams, 1962) and IgY, which is sometimes identified with this fraction (Hatta et al., 1990) |

**Table 3.** Concentration of Egg Immunoglobulins

| | IgY (IgG) mg/ml | IgM mg/ml | IgA mg/ml |
|---|---|---|---|
| Serum | approx. 6 | approx. 1.3 | approx. 0.6 |
| Yolk | approx. 25 | (<0.02) | (<0.03) |
| Clear part of egg | (<0.03) | approx. 0.15 | approx. 0.7 |

Numbers in brackets are to be taken as negative results.

The above value for the concentration of IgY in the yolk is rather high compared to values tallying with the concentration in the serum from some other researchers (Rammensee, 1984), but the disparity may be due to the wide daily fluctuation of the amount of IgY in the yolk (Schade et al., 1991, see Chapter 8).

The concentration of specific antibodies in the yolk are similar to the antibody-profile of the serum - a fact which is said to show that serum-IgG is channeled into the yolk during the follicle's ripening (Patterson et al., 1962).

To establish the presence of specific yolk IgY after an infection or immunisation, no IgY-extraction and purification processes are needed, as the yolk can be examined directly for antibodies without any manipulation (Sachsenweber et al., 1994; Gast et al., 1997; Schwarzkopf, 1994).

Likewise, to produce specific yolk-antibodies for therapeutic purposes, no yolk IgY extractions are needed (Erhard et al., 1996).

However, for research and diagnostic purposes at least roughly purified yolk IgY should be used.

### Separation of the IgY-containing water-soluble phase of egg yolk from water-insoluble phase

To isolate yolk IgY, various available processes may be used alone or in combination according to the following criteria: amount, purity and biological activity. The bother and cost involved are further relevant factors. Given a high level of antibody activity, an isolation-process entails a choice between amount and purity, since every step of purification principally involves a quantitative loss of yolk IgY.

All extraction processes begin by separating yolk and egg white. This should be done in such a way as to leave no contaminating traces of the egg white on the yolk. Various authors write of rinsing the yolk in flowing water or a buffer, and others roll yolks on filterpaper, to let the paper soak up the traces of the clear part. After removal of the vitellin membrane, the yolk is put into a measuring cylinder.

The following set of steps for IgY isolation is only a rough generalisation:

- Dilution of the yolk with distilled water or buffer or straightaway with an extraction medium, to bring the suspension into a precipitable or centrifugeable form; when the yolk is diluted with distilled water or buffer it is useful to keep the solution at -20 °C for 72 h followed by thawing (slowly at 4 °C) before further use
- Separation of the lipophile from the watery phase of the yolk (organic solvents, precipitation)
- Obtaining of IgY from the watery phase (pecipitation, chromatographic separation); the final IgY-extract of a selected extraction procedure should be dialysed against an appropriate buffer (0,1 % sodiumazide may be added as preservative)
- Isolation of specific IgY

To classify the various isolation procedures, the first step is usually definitive. The consequent purification processes vary widely in kind and combination and from institute to institute, so they are not suitable for classifying the various methods.

## Subprotocol 1
## Extraction Using Organic Solvents

By now there are many non-wasteful extraction procedures, so the waste-disposal of organic solvents, which often tended to be used for isolating IgY, should early be taken into consideration in choosing a suitable extraction procedure. A procedure which in this respect is rather extravagant has been described by Bade and Stegemann (1984).

## Outline

The IgY-extraction according to the chloroform-polyethylene glycol method (CH-PEG) is shown schematically in Figure 1.

## Materials

### Equipment

- cool-centrifuge (Jouan MR 1822)
- tubes (Falcon, Blue Max™ 50 ml, Propylene conical tube)
- pH-meter

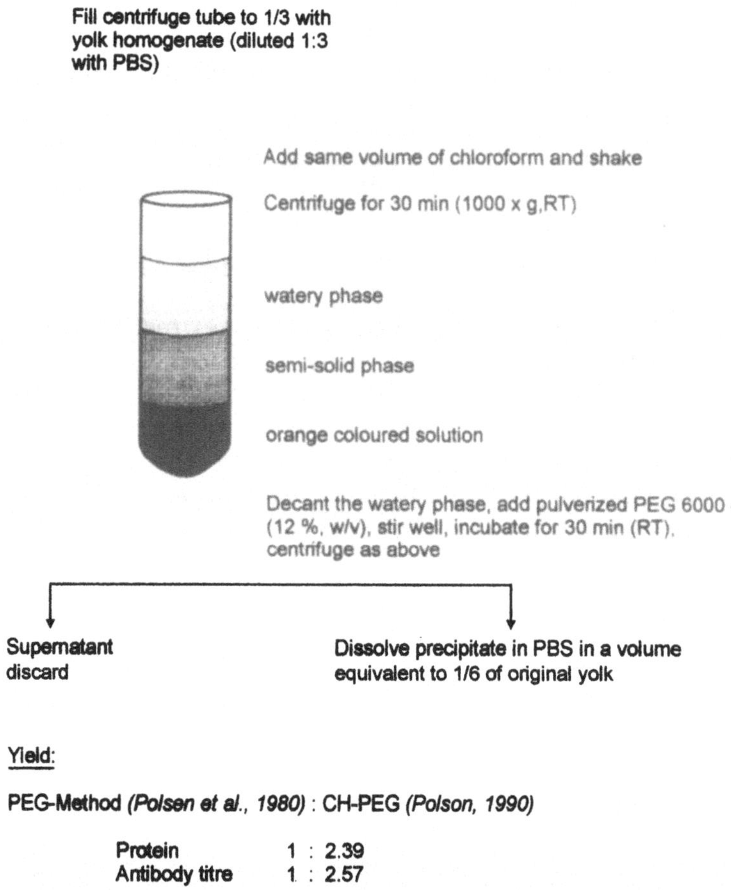


**Fig. 1.** IgY-extraction by the chloroform-polyethylene glycol method (CH-PEG).

**Reagents, solutions**

- Chloroform (p.a)
- PBS
- PEG (Mw 6000)

## Procedure

The following procedure has been described by Bade and Stegemann (1984).

1. A yolk is mixed with 100 ml of cold isopropanol.
2. The supernatant fluid is disposed of, and the process is thrice repeated with 100 ml of isopropanol each time.
3. The yolk is then washed twice with 100 ml of acetone each time.
4. After removal of the acetone, the sediment is dried on filter-paper and reconstituted in PBS. The yolk IgY can be further purified, if needed, through ion-exchange chromatography or through addition of polyethyleneglycol (PEG).

Chloroform has been used variously in the extraction of yolk IgY. There are a few reports of its being used alone rather than in combination with other procedures. IgY extracted with chloroform has not been used for the production of diagnostic reagents. Instead, chloroform extracted IgY served in a similar way like plain yolk as test material for measuring antibody titres after infection or immunization.

## Results

See Figure 1.

## Troubleshooting

The supernatant may not be clear but milky and dull. If so, it is advisable to shake it with chloroform again (in the same proportions by volume).

In investigating antibodies against *Rous sarcoma virus*, Aulisio and Shelokov (1967) used a chloroform extract. They mixed 1 volume of yolk with the same amount of physiological saline and 2 volumes of chloroform. The mixing and settling of the sediment followed at least four repeats of this procedure. After cooling overnight and centrifuging off at 2000 rpm, the clear watery supernatant was used for investigation in the indirect fluorescence-antibody test (IFAT). Mohammed et al. (1985) described a procedure which was similar but avoided the fourfold addition of chloroform during their investigations into *Mycoplasma gallisepticum* by ELISA.

## Subprotocol 2
## Polyethylene Glycol Precipitation (PEG) According to Polson et al. (1980)

The use of (PEG, MW 6000) for separation of yolk-IgY was introduced by Polson et al. (1980). This method is so widely accepted that it can be regarded as standard technique.

### Materials

**Equipment**

- cooling centrifuge
- refrigerator
- tubes
- funnel
- filter-paper
- pH-meter

**Reagents, solutions**

- PBS
- PEG (pulverized, MW 6,000)

## Procedure

The IgY-extraction according to the polyethylene glycol precipitation method is shown schematically in Figure 2.

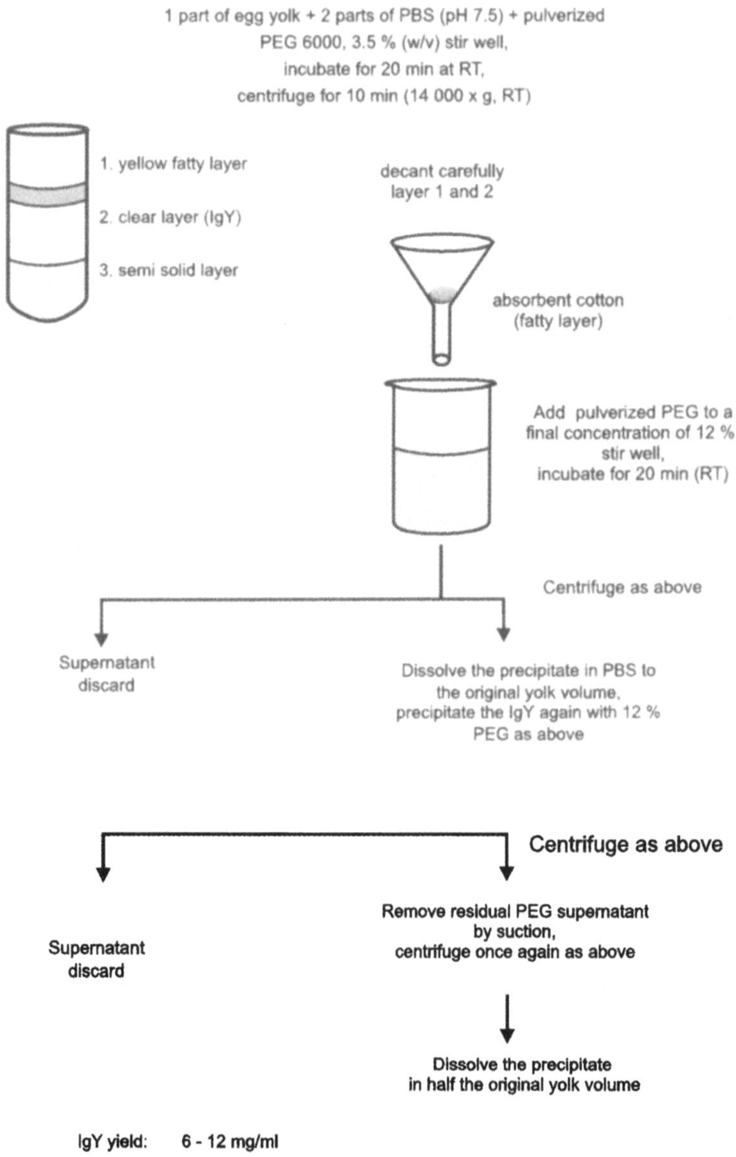


**Fig. 2.** IgY-extraction by polyethylene-glycol precipitation (PEG).

## Results

See Figure 2 and Tables 6 and 7.

## Troubleshooting

If the purity of the product is deemed inadequate, the addition of 2 M ammonium sulphate leads to precipitation of protein, while PEG is removed with the supernatant fluid. After completion of the IgY-extraction procedure, the final product should be dialysed against PBS or any other appropriate buffer. See Chapter 8 for a very detailed description of the procedure.

# Subprotocol 3
# Polyethylene Glycol/Ethanol Precipitation (PEG/Alc) According to Polson et al. (1985)

Comparative investigations with other methods revealed that the IgY-yield of this original PEG-method is relatively low (Akita and Nakai, 1993; Schwarzkopf, 1994). Furthermore, residues of PEG may disturb some immunological assays (osmotic potency of PEG).

Consequently Polson et al. (1985) presented an improved method, in which a protein precipitation step with cold ethanol was added. In this way, residual PEG was completely removed. Since some yolk substances are solved better in organic solvent than in water, such contaminants are removed by this method.

## Materials

See Subprotocol 2, additionally ethanol, refrigerator (-5 °C/ -20 °C).

## Procedure

The IgY-extraction according to the polyethylene glycol/ethanol precipitation method is shown schematically in Figure 3.

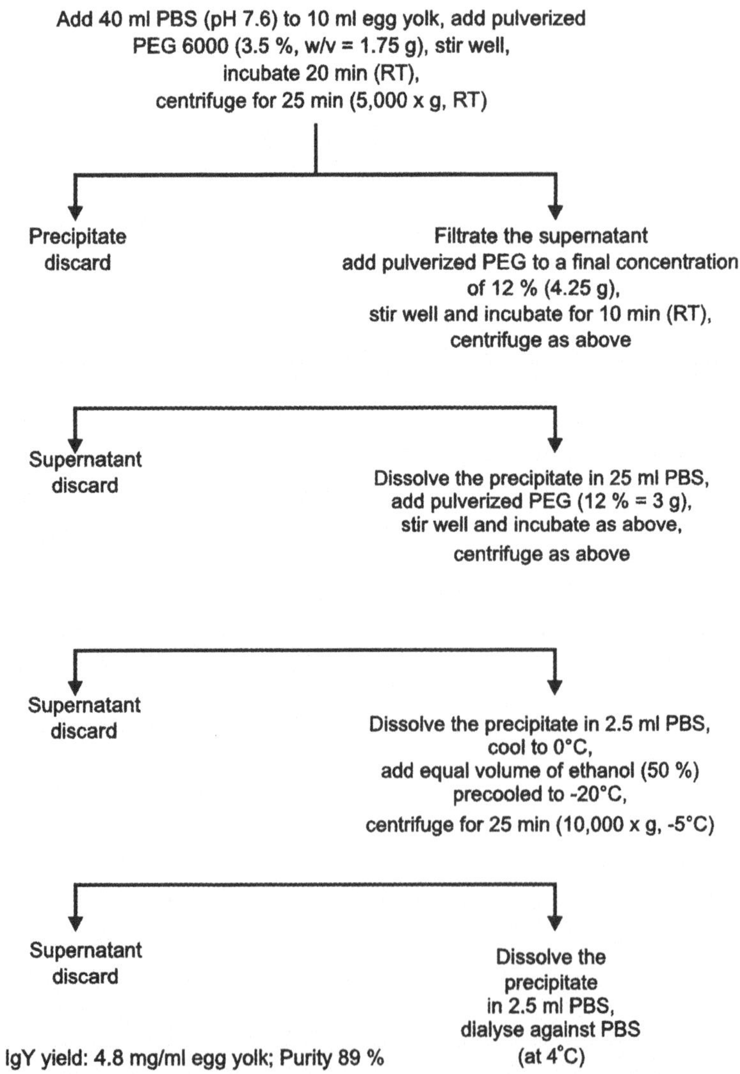


**Fig. 3.** IgY-extraction by polyethylene-glycol/ethanol precipitation (PEG/alc).

## Results

See Figure 3 and Tables 6 and 7.

## Troubleshooting

A combination of chloroform extraction with the PEG-method, which Polson presented in 1990 produced a 2.6-fold greater activity and a 2.4-fold greater yield of protein. The steps of this procedure - known as the chloroform-PEG-method - are shown in Figure 1.

If further purification of the IgY-preparation from PEG is needed, the distribution-coefficient of PEG in water:chloroform of 1:20 can be used, which means that a further shaking of the preparation with chloroform may lead to a notable reduction of the PEG-concentration.

# Subprotocol 4
# Dextransulphate Precipitation (DS) According to Jensenius et al. (1981)

The precipitation of lipoprotein from serum by dextransulphate (Burstein and Pavermann, quoted by Jensenius et al., 1981) was used by Jensenius et al. (1981) to separate yolk lipid from the watery phase of yolk to obtain yolk-IgY. This procedure takes place at room temperature.

## Materials

**Equipment**

- cooling-centrifuge
- tubes
- pH-meter

**Reagents, solutions**

- TBS
- dextran sulfate
- $CaCl_2$

## Procedure

The IgY-extraction according to the dextransulphate precipitation method is shown schematically in Figure 4.

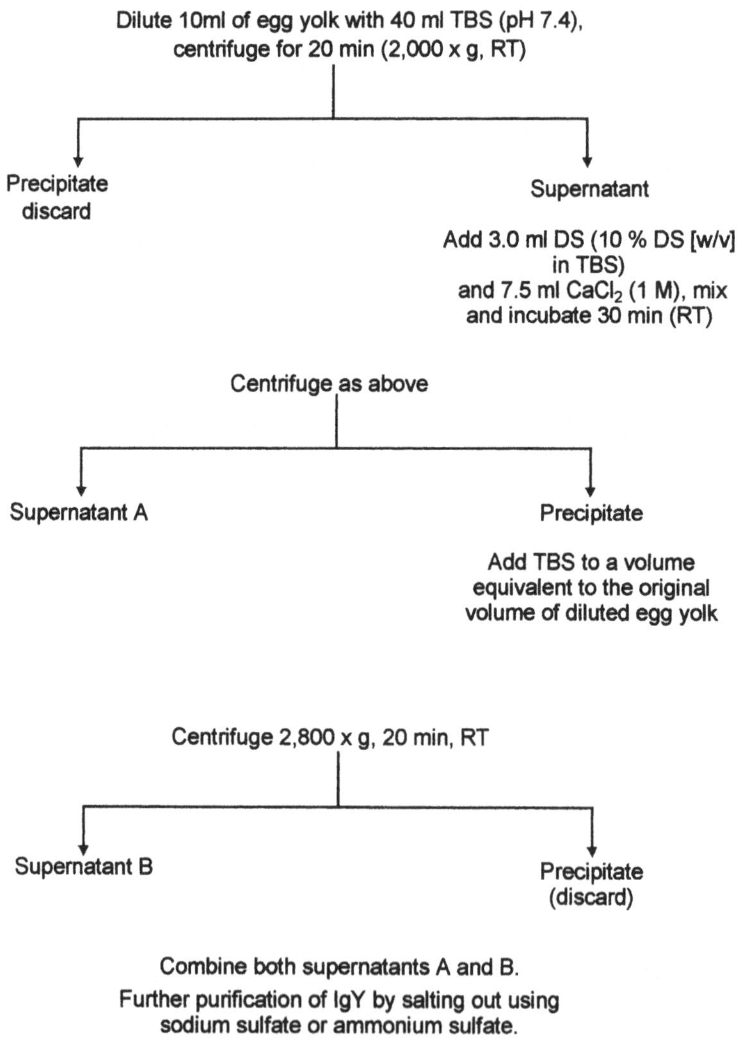


**Fig. 4.** The diagram shows IgY-extraction by means of dextran sulphate precipitation (DS).

## Results

See Tables 6 and 7.

## Troubleshooting

This method is flexible insofar as the first step of isolation can be repeated if there is any cloudiness in the first supernatant fluid due to a too meagre addition of DS and $CaCl_2$.

Yolk-IgY in the supernatant fluid can finally be precipitated by addition of either sodium sulphate or ammonium sulphate according to "salting out procedures" (see below). This procedure isolates about 70-80 % of the yolk-IgY.

# Subprotocol 5
# Water-Dilution Method (WD)

With a neutral pH and a low concentration of ions, the yolk lipids aggregate. Low pH-values also lead to a separation of IgY (Jensenius et al., 1981; Akita and Nakai, 1992). This aggregation of lipids is used in the water-dilution method to separate IgY.

The method is used in so many forms that none can be viewed as standard. As an example, the water dilution procedures described by Akita and Nakai (1993) and by Schwarzkopf (1994) are given.

## Materials

**Equipment**

- cooling centrifuge
- refrigerator
- deep freezer
- tubes
- pH-meter

**Reagents, solutions**

- distilled water
- sodium sulphate
- ammonium sulphate

## Procedure

The IgY-extraction according to the water-dilution precipitation method (Akita and Nakai 1993) is shown schematically in Figure 5.

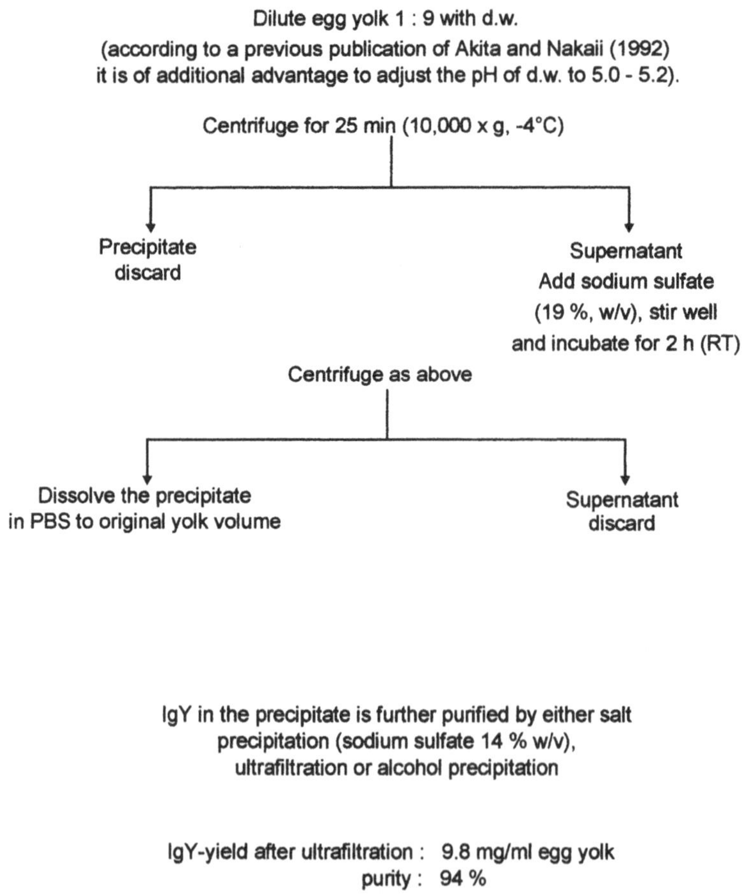


**Fig. 5.** IgY-extraction by the water dilution method (WD) according to Akita and Nakai (1993).

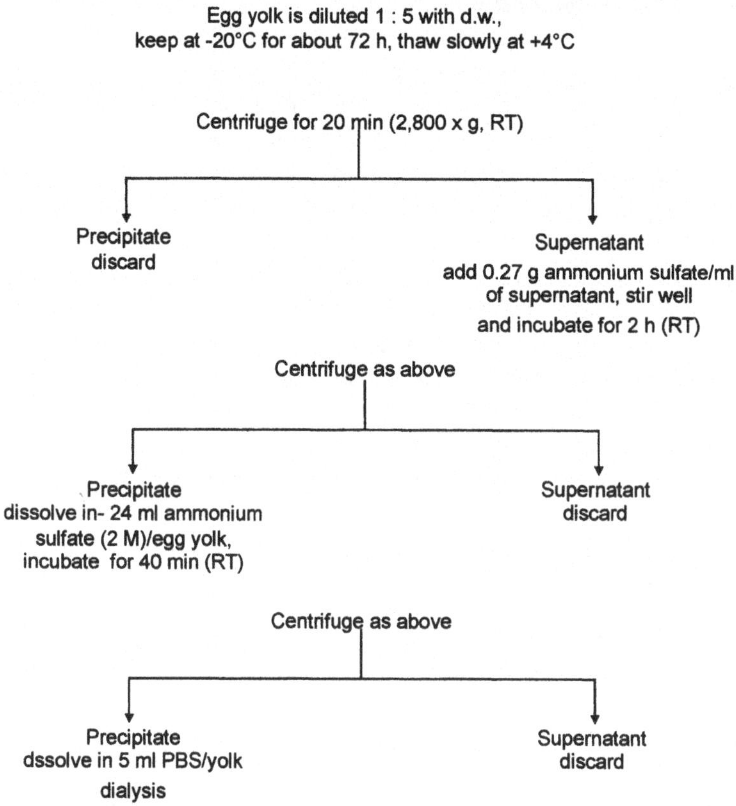


**Fig. 6.** IgY-extraction by the water dilution method (WD) according to Wallmann et al. (1990).

The water dilution procedure according to Wallmann and Staak described by Schwarzkopf (1994) is shown schematically in Figure 6.

## Results

See Figures 5 and 6 and Tables 6 and 7.

## Troubleshooting

Jensenius et al. (1981) used a ratio of yolk:water of 1:10, froze the mixture to -20 °C and thawed out, centrifuged off and worked with the supernatant fluid. Wallmann et al. (1990), as also Schwarzkopf (1994), used a ratio of yolk:distilled water of 1:5, likewise froze the mixture to -20 °C and especially emphasised that the mixture should be thawed slowly at 4 °C.

Yokoyama et al. (1993) investigated the effects of varying the ratio of distilled water added to yolk from 1:4 to 1:10 and of incubating the mixtures at temperatures between 4 °C and 15 °C. They also examined the influence of various amounts of hydroxypropylmethylcellulose phthalate (HPMCP) as a lipid-precipitating active substance.

The authors found the following parameters to be optimal in yielding IgY:

- dilution of yolk:water 1:8
- addition of HPMCP 0.25% (v/v)
- working temperature 10 °C

## Subprotocol 6
## Further Procedures for Separating Lipoproteins/Lipids from Water-Soluble Proteins

The following procedures for separating lipoproteins/lipids from water-soluble proteins are less widely adopted. Thus, these procedures are not described in detail.

## Procedure

### Lipid-precipitating qualities of "natural gums"

Hatta et al. (1990) examined 12 various 'natural gums' as regards their lipid-precipitating qualities.

10 ml of yolk were diluted with 10 ml of distilled water, then 40 ml of distilled water were added together with 60 ml of the

respective gum and incubated at room temperature for 30 minutes. Finally the mixture was centrifuged at 10,000 x g for 15 minutes. The watery phase of the supernatant fluid was examined for lipid contents. The lipids were measured by shaking the mixture with chloroform-methanol (3:1) and evaporating and weighing it (see above, Yokoyama et al., 1993). The natural gums with the greatest yield of IgY proved to be 3 various preparations of carrageenan with 0.4-1.1% and xanthan gum with 0.7% residual lipids.

### Combined octane-acid/ammonium sulphate precipitation

The combined octane-acid/ammonium sulphate precipitation introduced by McKinney and Parkinson (1987) for the simple isolation of mammalian IgG from serum was used by Schwarzkopf (1994) to obtain yolk-IgY:

1. Dilute yolk 1:5 with 0.06 M acetate-buffer at pH 4.0.
2. Add 25 µl of octane acid per ml of the mixture.
3. Incubate for 30 minutes at RT and centrifuge (10,000 x g, RT).
4. Add PBS to the supernatant (tenfold, v:v), final pH 7,2-7,4.
5. Add 10 volumes of 45% ammonium sulphate (AmS) to 9 volumes of the PBS/supernatant, incubate the mixture for 30 minutes at 4 °C and centrifuge (10.000 x g, 4 °C).
6. Add 5 ml PBS/yolk followed by dialysis against PBS.
7. The dialysate was heated to 50-55 °C for 20 minutes and finally purified by being centrifuged (5.000 x g) and filtered (0.45 µm).

### Separation of the IgY-fraction from the aqueous phase of egg yolk

The procedures described above are ways of separating the lipophile and watery phases of the yolk. Purification steps were mentioned insofar as they directly belong to a technique presented.

The result of the first step of the procedure is isolation of the watery phase with yolk-IgY but also other proteins, especially alpha- and beta-livetin. The second step involves procedures suited to isolate IgY from the aqueous phase of egg yolk.

## Subprotocol 7
## Precipitation of Yolk-Protein With Ammonium Sulphate [$(NH_4)_2SO_4$ - 'AmS'] or Sodium Sulphate [$Na_2SO_4$ - 'NaS']

Proteins can be precipitated with AmS or with NaS* whatever procedures have been used till then. (* For historical reasons the concentration of AmS is given in percent saturation and of NaS in weight/volume.)

Steps taken for the salt-precipitation of proteins are in the following sequence: A slow addition of AmS or NaS in a measured amount as a substance or in a concentrated solution. Thorough mixing of the substances is important, to avoid having an uneven concentration of the salt. Since the precipitation of certain proteins from a mixture by AmS or NaS depends on the concentration of the salt, this procedure can be used not only for isolating desired proteins but also for eliminating undesired ones (see below, Orlans and Rose, 1972, as also Yamamoto et al., 1975). There follows a phase of incubation at room temperature under stirring. A precipitate is obtained through centrifuging, taken up in a certain amount of a desired buffer or distilled water and finally dialysed.

### Outline

This protocol gives information about a lot of IgY-precipitation methods not described in detail. A procedure of IgY-precipitation by AmS is exemplary shown in Subprotocol 5.

### Materials

See Subprotocol 5.

## Procedure

Orlans and Rose (1972), as also Yamamoto et al. (1975), used AmS precipitation for preliminary purification by homogenising yolk in 4 volumes of 16% saturated AmS and by isolating the watery supernatant fluid through incubation and centrifuging (see Table 1, Chapter 8). From this fluid, IgY was obtained as a precipitate in 50% saturated AmS and purified through a sequence of three further precipitations with 40% and twice 37% AmS-saturation. Pure IgY was obtained from the dialysed product by gel-filtration via sephadex G 200. Similar procedures were used by Kuroki et al. (1993), as also by Loeken and Roth (1983). They diluted the yolk with an equal volume of PBS and centrifuged at high velocity. IgY was obtained from the supernatant fluid by using AmS in three steps of precipitation, once with 50% AmS-saturation and twice with 33% saturation.

Several steps of precipitation are often described. Wallmann et al. (1990) used three steps of AmS-precipitation, with addition of solid AmS at each step to restore 33% saturation. Ntakarutimana et al. (1992) obtained their material through chloroform extraction. They used a gradual addition of 20% (w/v) NaS to yield IgY.

Hatta et al. (1990) used natural gum for the first step of isolation. To obtain IgY, NaS was added to the centrifuged supernatant fluid to a concentration of 0.2 M.

As a further purification after treatment with dextran sulphate, Jensenius et al. (1981) advised adding 20g NaS to 100 ml supernatant fluid and a final reprecipitation by adding 6.2 ml of 36% (w/v) NaS to 10 ml. Schmidt et al. (1989) took the advice but used AmS in a final saturation of 50%.

Polson et al. (1980) likewise saw a further possibility of purification of IgY after PEG-preparation in the precipitation of protein after addition of 2 M AmS.

A systematic investigation into the suitability of various AmS concentrations was presented by Svendsen et al. (1995). The initial material was supernatant fluid obtained by the method of water-dilution (yolk:distilled water 1:10) and freezing and thawing.

**Table 4.** Relation Between IgY-Yield and AmS Concentration

| AmS Concentration (%) | Yield of IgY (%) |
|---|---|
| 01 - 10 | 0 |
| 11 - 20 | 0 |
| 21 - 25 | 0 |
| 26 - 30 | 29 |
| 31 - 40 | 29 |
| 41 - 50 | 3 |

In using AmS the authors recommend the following procedure:

1. Diluting the yolk 1:10 with distilled water.
2. Freezing and thawing the yolk-solution.
3. Centrifuging and filtering the supernatant fluid.
4. Adding AmS substantially up to 25% saturation, and discarding the precipitate after centrifuging.
5. Adding AmS as salt calculated for 40% saturation, and keeping the precipitate after centrifuging.
6. Taking up the precipitate in distilled water or buffer.
7. Dialysing against distilled water or buffer.

## Results

See Table 4.

## Troubleshooting

See Subprotocol 5.

## Subprotocol 8
## Gelfiltration of an IgY-Extract According to the Subprotocols 1-7

IgY has a molecular weight of 165,000 (Gottstein and Hemmler, 1985). Other reports vary between 165 and 220 kDa (see Fischer, 1996; Fischer et al., 1996).

The separation of proteins according to their molecular weight through gelfiltration is a further stage of purification of yolk IgY but rarely reported except in the desalting of preparations through sephadex G 25 instead of through a tedious dialysis.

### Materials

**Equipment**

- column (2 x 90 cm) and accessories:
  - valve
  - connecting piece
  - tubes
  - pump, like peristaltic pump P1, Pharmacia
  - UV-monitor
  - recording-unit (plotter or printer)
  - equipment for volume-reduction through vacuum dialysis (collodium tubes, Sartorius) or through lyophilisation

**Reagents, solutions**

- watery solution containing IgY
- buffer (e.g.PBS, containing 0,5 M NaCl to avoid non-specific binding)
- Sephacryl S-300 (Pharmacia)
- Blue-Dextran 2,000 (Pharmacia, MW 2,000 kDa)

If no specially made columns are available, one can proceed as follows (more detailed descriptions are given in the Pharmacia booklet 'Gel Chromatography').

## Procedure

1. Wash and transfer Sephacryl S-300 to PBS, choosing a gel-viscosity which enables air-bubbles to rise.
2. Put several ml of PBS into a column (close the lower end of the column).
3. Transfer the gel into the column carefully, the best way being to pour it down a glass rod.
4. Let the gel set, then connect the column to a buffer-container. It is advisable to equilibrate the columns with several column-volumes of buffer. Separation occurs under a pressure depending on the difference in level between the lower end of the column and the buffer-level in the container. It is simpler to use a pump.
5. After equilibration of the column, apply the mixture of proteins to be separated (e.g.5 ml, depending on the column size). It is useful to connect a corresponding sample-container via a valve to the upper end of the column and to mark the sample with blue-dextransulphate, to see if the column separates properly.
6. Collect the fractions as soon as the blue fraction is eluted from the column, since blue-dextran 2000 has a higher MW than the proteins to be separated.
7. Examine the IgY-content of the samples collected with a suitable assay e.g. immuno- electrophoresis (see Figure 7) or ELISA (see Chapter 5).
8. Pool and concentrate fractions containing IgY, for instance by vacuum dialysis.

## Results

For results see Figure 7.

Hassl and Aspöck (1987) produced yolk IgY-antibodies against *Toxoplasma gondii* and used the PEG method according to Polson et al. (1980) for isolation. For purifying the yolk-IgY, the authors set a sephacryl S-300 column (1.6 x 60 cm), to which they applied 2 ml of the IgY fraction dissolved in PBS.

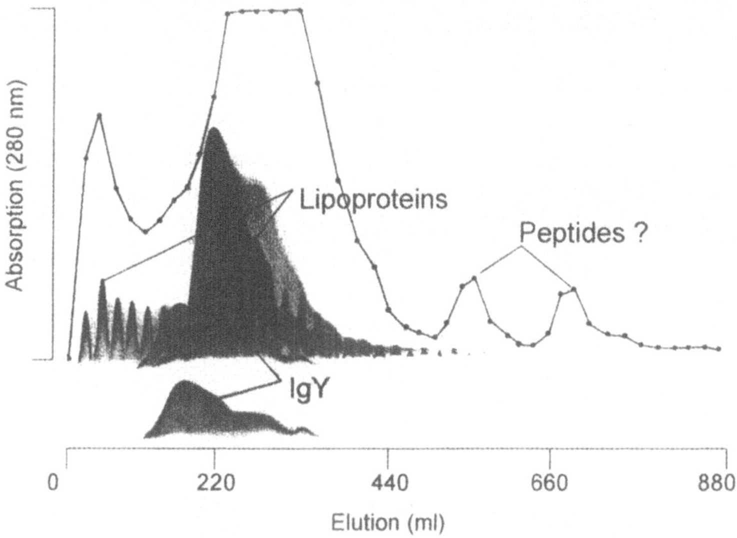


**Fig. 7.** The elution-profile of a yolk-extract (yolk:distilled water 1:10 frozen, thawed out and centrifuged) according to gel-filtration using Sephacryl S300. The continuous line marks the absorption at wavelength of 280 nm. For the development of immuno-electrophoresis (fused rocket-immunophoresis, see Chapter 5) shown in the upper part of the figure, a polyspecific anti-hen serum antibody was used. The profile in the lower part of the figure was developed with a monospecific anti-IgY antibody. From the gathered fractions, IgY can be isolated by means of a precipitation procedure.

In the eluate, two protein fractions were isolated, of which the lighter one showed antibody activity and, after reduction to the volume applied, yielded the initial titre of the sample used.

A comparison between the following methods - isopropanol (Bade and Stegemann, 1984), PEG (Polson et al., 1980) and PEG ethanol (Polson et al., 1985) - revealed three peaks in each of the first two cases and one peak with a small 'shoulder' in the third. This was taken as evidence of the superiority of the PEG-ethanol procedure over the two others in purifying IgY.

In a further investigation (Hassl and Aspöck, 1988) the authors used a two step procedure (hydrophobic interaction chromatography in combination with gelfiltration) for purification of IgY.

## Troubleshooting

Mistakes may be made in pouring gel into the columns. If it is poured unevenly, eluated fractions may be not sufficiently uniform, and if the gel has not the appropriate consistence, air-bubbles will be caught. The pressure applied should depend on the gel's properties. Too much pressure on sensitive gels compresses them and changes their separation-properties. Likewise the elution's speed influences the efficacy of separation, since too high a speed leads to poor separation.

## Subprotocol 9
## Ion Exchange Chromatography of an IgY-Extract According to the Subprotocols 1-7

Ion-exchange chromatography is a widespread method for separating proteins from mixtures and is also used for isolating immunoglobulins. Proteins are bound electrostatically to an ion-exchange matrix with a reverse-charge. According to the strength of binding, the proteins are pushed away - i.e. detached - by increasing concentrations of ions of the buffer.

The second method of detaching them relies on changing the pH till the iso-electrical point of the respective protein is reached, whereupon the protein becomes free of charge and no longer binds to the matrix.

## Materials

**Equipment**

- pH-meter
- shaker
- homogeniser (e.g. Ultra-Turrax)
- refrigerator (-20 °C)
- collodium-tubes (Sartorius)
- water-jet-pump (vacuum-dialysis)
- centrifuge

**Reagents, solutions**

- Bakerbond wide-pore prepscale Abx, 40 µm (J.T.Baker)
- MES-buffer (25 mM, pH 5,6)
- phosphate buffer ($H_2PO_4$, pH 7,0)
- ammoniumsulphate

## Procedure

The course of an ion exchange procedure is shown schematically in Figure 8. In this case a so-called batch procedure is described.

## Results

Under the chosen conditions, the yield is 70 - 80 % and the purity 80 - 90 %.

Hatta et al. (1990) used as their raw material for ion exchange chromatography the supernatant fluid obtained by lambda-carrageenan treatment, to which they added solid $Na_2PO_4$ to a concentration of 0.02 M, using 3 N HCl to bring the pH to 8.0. The previous equilibration of the DEAE-sephacel-column, as also the washing (20 column-volumes), after production of the supernatal fluid, was performed with the same buffer. Bound protein was finally detached and eluated with a 0.2 M phosphate buffer.

Likewise Gassmann et al. (1990) used a rising ion-gradient. They began absorption with a 0.015 M potassium phosphate buffer (pH 8) and raised the ion concentration to 0.2 M at the same pH.

Schade et al. (1992) used a batch procedure with application of the DEAE ion exchanger (Pharmacia). Their starting material was 1:10 diluted yolk after freezing, thawing and filtering. The filtrate was dialysed against a 0.05 M phosphate buffer at pH 6.8 and finally mixed with the same buffer of swollen and equilibrated DEAE-A 50. The mixture was repeatedly washed. After the molarity of the buffer was raised, yolk-IgY was detached and obtained in the supernatant fluid or through filtration in the filtrate.

Egg yolk extract:
Dilute one egg yolk in 80ml MES-buffer (0.01 M, pH 5.6), homogenise or shake, keep at -20°C for 24 h, thaw slowly and centrifuge or filtrate

Ion exchange gel:
wash 1g ABx thrice with 50 ml MES buffer, 25 mM (the pH should be stable at 5.6)

Mix the extract with the equilibrated ABx, shake for 30 min at room temperature, allow to settle and discard the supernatant, wash with 80 ml equilibration buffer,

↓

Desorb the bound IgY Ab by adding 40 ml elution buffer (0.2 M $H_2PO_4$, pH 7.0), shake for 30 min, allow to settle and decant the IgY-containing supernatant

↓

Concentrate the supernatant (e.g. by vakuum dialysis and refrigerate) or Lyophilise the supernatant

ABx = Bakerbond Wide-Pore Prescale ABx, 40 µm, J.T. Baker
MES = Morpholinoethane sulfonic acid p.a. SERVA

**Fig. 8.** IgY-extraction by an ion exchange procedure (batch procedure).

## Troubleshooting

It is important to keep exactly the given pH values and salt concentrations and to allow the IgY to bind to the matrix in an appropriate time.

## Subprotocol 10
## Affinity Chromatography of an IgY-Extract According to the Subprotocols 1-7

The principle of affinity chromatography (AfC) consists in immobilising a reaction-partner (ligand) on a matrix. The specific second reaction partner in a mixture couples with the ligand, whereas the non-specific substances are rinsed away. The bound second reaction partner is finally detached through certain manipulations and isolated. In IgY-technology as with other fields, two principles of use are especially interesting:

- The obtaining of IgG-isotypes via AfC-columns with immobilised protein A (from *Staphylococcus aureus*) or protein G (from *Streptococci* of the C- or G-group), to which IgG-isotypes from various mammalian species bind differently. IgY from hens binds neither with protein A nor with protein G, in contrast to IgG from ducks, which couples with protein A (Higgins et al., 1995). Kronvall et al. (1970) report on the binding of IgG from rheiform birds' serum as a further exception. Therefore it is not possible to mark chicken IgY via enzyme-marked protein A or G for use in ELISA, or via FITC-marked protein A or G for use in fluorescence immuno histochemistry.
- The situation is quite different as regards immobilising a specific antigen on the matrix. The IgY antibodies to be obtained bind to the corresponding antigen and can be detached and eluted, as with IgG from mammals, by lowering the pH value to between 2.3 and 2.5.

The AfC procedure is often used. Since it is not essential to start with extremely purified IgY, a number of preparatory steps can be omitted. The final product is selectively obtained antibody, whose specificity depends on the composition of the immobilised antigen.

Thus AfC blunts the occasional demand for use of SPF animals. AfC is reversible, since in immobilising a specific antibody as a ligand on the matrix, the corresponding antigen is bound then isolated. Apart from it's universal applicability, this procedure is practical in revealing contamination with penicillin (De Leuw, 1996), in purifying parathyroid related protein (1 -

36) (Rosol et al., 1993) and in obtaining pure human alpha$_2$-antiplasmin (Lee et al., 1997). As a matrix for serological work, the material most often used is sepharose 4B after activation with CNBr ('CNBr activated sepharose 4B').

CNBr activated sepharose 4B is available commercially, otherwise instructions for use may be taken from Tijssen (1985).

## Materials

**Equipment**

See also Subprotocol 8.

- column of appropriate size (small size)
- shaker
- refrigerator

**Reagents, solutions**

- CNBr-activated Sepharose
- HCl (1mM)
- $NaHCO_3$-buffer (0.1 M, pH 8.3, 0.5 M NaCl)
- ethanolamin or 0.2 M glycin (pH 8)
- acetate buffer (0.1 M, pH 4 with 0.15 M NaCl)
- Glycin/HCl (pH 2.4)
- PBS
- Tris (3M)

## Procedure

**Coupling of the ligand**

Coupling of the ligand according to the second principle as described in the introduction (antigen as ligand to "fish" the IgY) with CNBr activated sepharose B4 is performed in the following way (see also booklet 'Affinity Chromatography - principles and methods', Pharmacia, LKB, Biotechnology):

1. Weighing the calculated amount of lyophilised CNBr activated sepharose 4B (1 g lyophilisat produces about 3.5 ml of swollen gel).

2. Washing and swelling on a sintered glass filter (G3) with 1 mM HCl (need 200 ml/g sepharose).
3. Dissolving the protein for coupling in a coupling-buffer ($NaHCO_3$-buffer, 0.1 M, pH 8.3 with 0.5 M NaCl, ca. 5-10 mg protein/ml gel).
4. Mixing the protein/gel end over end (2 h RT or overnight at 4 °C, no magnetic stirrer).
5. Blocking remaining active groups by transferring the gel to a coupling buffer with a blocking substance (1 M ethanolamin or 0.2 M glycin, pH 8, 2 h at RT or 16 h at 4 °C).
6. Rinsing to remove superfluous absorbed protein (coupling-buffer followed by acetate buffer (0.1 M, pH 4 with 0.15 M NaCl), followed by coupling-buffer.
7. Rinsing to remove blocking substances (coupling-buffer).
   The protein-sepharose-conjugate is hereby usable in AfC.

Conservation at 4 - 8 °C after adding $NaN_3$ in a final concentration of 0.1%.

**Isolation of specific IgY**

Isolation of specific IgY by binding to the corresponding antigen immobilised on CNBr-activated Sepharose:

1. Equilibration of the columns with PBS, application of the IgY solution, rinsing with PBS (low flow rate! Non-bound material must be wholly removed i.e. the absorption measured by the UV-monitor should be <0.1).
2. Elution (addition of the elution-buffer [glycin/HCl, pH 2.4] till no protein can be detected in the eluate).
3. Adjusting the pH value of the eluate with 3M Tris to 7.2.
4. Dialysis against a chosen buffer (24 h, 4 °C).
5. Centrifugation (2.800 x g, 15 min).
6. Concentration in the dialysis tube, e.g. by Carboxymethyl-cellulose (Aquacide$^R$).

Store the specific IgY in portions at -20 °C and the columns after long rinsing with PBS and addition of 0.1 % $NaN_3$ at 4 °C. The columns can be used repeatedly.

## Results

An alternative matrix was presented by Staak et al. in 1996. The authors used pulverised micro-titre-plates to which the ligand was bound as in ELISA. Affinity chromatography followed the above method. The yield is notably less than in the case of chromatography with CNBr activated sepharose 4B, but the method is cheap and does not rely on imports. These affinity columns, too, can be used several times.

## Troubleshooting

It is important to avoid overloading of the matrix with IgY (see Procedure).

# Subprotocol 11
# Amount of Specific Ab Extractable From Egg Yolk of Immunised Hens

In this protocol it is shown, that the amount of specific Ab obtained by affinity chromatography depends obviously on the antigen used for immunisation. In this respect, it is the same as with the immunisation of mammals.

## Materials

As with Subprotocol 3, Chapter 3.

## Procedure

As with Subprotocol 3, Chapter 3.

After repeated immunisation with several antigens, antigen-specific IgY from the total fraction of yolk IgY is purified through affinity chromatography, then the percentage of specific antibodies is found (see Table 5).

## Results

The results are shown in Table 5.

**Table 5.** Percentage of Specific Antibodies in the Whole Egg Yolk of Chickens Immunised With Different Antigens

| Antigen | Specific IgY (%) |
|---|---|
| Escherichia coli K88 fimbrial antigen | 0.9 - 1.8 |
| Recombinant bovine somatotropin | 2.7 - 8.0 |
| Human IgG | 7.4 - 18.9 |
| Low density lipoprotein | 2.1 - 5.6 |
| Bovine IgG | 12.4 - 13.5 |
| Specific IgY range from 1% to 20% | |

## Troubleshooting

As with Subprotocol 1, Chapter 3.

## Comments

### Comparison of IgY-isolation procedures as regards yield and purity

Three criteria serve as a basis for comparing various procedures for isolating IgY from the yolk of immunised hens:

- Yield of IgY
- Purity of IgY isolated
- Antibody activity i.e. the amount of specific yolk-IgY.

Insofar as they are mentioned in the literature, the measurement procedures for the single criteria are given as follows. For implementing serological standard-techniques, reference is made to the relevant literature.

The yield of IgY without taking antibody-activity into account is expressed as mg/ml of the final product or as mg/yolk. These values are valid only if at the same time the purity of the yolk-IgY or its concentration as compared to the competing proteins is given.

For measuring proteins the following procedures were used: the Biuret method, the method according to Bradford (BioRad), Lowry's method, UV-absorption at 280 nm and the micro-Kjeldahl procedure (for instruction see Holtzhauer, 1997).

Immunological procedures for measuring the quantity and purity of the product included single radial immuno-diffusion (SRID, 'Mancini'-technique), used by researchers like Akita and Nakai, (1993), and the rocket -immuno electrophoresis or Laurell-technique, used by researchers like Schade et al., (1991); Svendson et al., (1995); Gross and Speck, (1996).

For testing the purity of isolated IgY, use was made of SDS polyacrylamide-gel-electrophoresis (PAGE) under non-reducing and reducing conditions.

Determination of antibody activities was generally performed with the test procedure for which the antibody was produced:

Enzyme-linked immunosorbent assay (ELISA) as a non-competitive procedure (e.g. Schwarzkopf, 1994) and competitive (C-ELISA) procedure (Kierek-Jazcuk et al., 1997; Gutierrez and Gurriero, 1991). Of all procedures, ELISA was the one most often mentioned.

The Agargel-immuno-double diffusion test (AGIDT, Ouchterlony) was used for instance by Yamamoto et al. (1975) and by Polson et al. (1980, 1990).

Also mentioned are the virus neutralisation test (VNT) (Czerny and Eichhorn, 1989), the indirect haemagglutination test (IHA) (Hassl and Aspöck, 1987), the radio-immuno-assay (RIA) (Rosol et al., 1993) and immuno-fluorescence-technique (FAT, IFAT) (Schmidt et al., 1989; Gervelmeyer, 1997).

Moreover, affinity chromatografical investigations enable yolk-IgY to be expressed in mg/ml. (Gassmann et al., 1990; Schwarzkopf, 1994). For details see Chapter 5.

The incomplete listing of investigation procedures shows how widely yolk-IgY is used.

### Comparison of various IgY-isolation techniques in regard to yield, purity and biological activity

Schwarzkopf (1994) compared various procedures for isolating IgY. The author used as starting material several yolks from the

**Table 6.** Comparison of Efficacy of Several IgY-Isolation Procedures Concerning Yield and Purity

| Procedure | Total protein content (TPC) of the extraction volume (= 6 ml) | Total IgY content of the extraction volume | mg of specific IgY anti-IgG (dog) | Titre in ELISA after extraction before chromatography |
|---|---|---|---|---|
| 1. WD | 684 mg/6 ml | 445 mg | 9.7 | 1:128.000 |
| | 11.4 % | 65 % | 2.2 % | |
| 2. DxS | 282 mg/ 6 ml | 278 mg | 6.0 | 1: 8.000 |
| | 4.7 % | 98 % | 2.2 % | |
| 3. PEG | 42 mg/6 ml | 32 mg | 0.8 | 1: 2.000 |
| | 0.7 % | 77 % | 1.0 % | |
| 4. PEG/Ethanol | 258 mg/6 ml | 246 mg | 5.4 | 1: 32.000 |
| | 4.3 % | 96 % | 2.2 % | |

Measurement of the total protein content (TPC) with the Biuret method

Identification and quantification of IgY with cellulose-acetate-electrophoresis and the Biuret method (total IgY content)

Isolation of specific IgY (total specific IgY content) through affinity chromatography, determination of protein with the Bradford method (BioRad)

Determination of titre in ELISA (Ab-titre after extraction)

eggs of a hen immunised against dog-Ig, and for each of the isolation procedures listed in Table 6, she used 40 ml of the evenly mixed yolk. The isolates were adjusted to a final volume of 6 ml, so that protein content and antibody activity could be directly compared with each other.

Extraction procedures:

- Water dilution method (Schwarzkopf)
- Dextransulphate precipitation (Jensenius et al., 1981)
- PEG precipitation (Polson et al., 1980)
- PEG and ethanol (Polson et al., 1985)

In a further comparison between various yolk-IgY isolation procedures, Akita and Nakai (1993) used a yolk mixture of eggs with antibodies against enterotoxic *Escherichia coli* (ETEC), dividing it into 4 portions of 20 ml as starting material for 4 different extraction procedures. There was evaluation not only of the final but also of the intermediate product (supernatant-SN) after the first separation of the lipophile and hydrophile phases.

**Water dilution method (WD):**

- Dilution of yolk:d.w. 1:9, 6 hrs incubation at 4 °C, centrifuging off 10.000 x g for 25 min at 4 °C
- Supernatant fluid + 19% sodium sulphate (WD-SN)
- Dissolving of the precipitate
- Ultra-filtration (WD-UF)

**Polyethyleneglycol (PEG/ethanol method according to Polson et al., 1985):**
The supernatant fluid after the first PEG precipitation is called PEG-SN, and the final product is called PEG-Alc.

**Dextransulphate precipitation (DS) according to Jensenius et al. (1981):**
The supernatant fluid after the first DS-precipitation as well as the 2nd supernatant fluid after elution of the precipitate is called DS-SN, and the final product after sodium-sulphate precipitation is called DS-As2.

**Xanthan method (Xan) after Hatta et al. (1990):**
The supernatant fluid after xanthan precipitation is called Xan-SN, and the final product after the 2nd xanthan precipitation and the following sodium-sulphate one is called Xan-As2.

The SN-preparations as well as the final product are compared in Table 7.

Antibody activity was measured in the indirect ELISA with antigen from E.coli. Both the 4 SN fractions and the 4 final products were investigated after having been uniformly adjusted to 0.5 μg IgY/ml.

The levels of extinction of the SN-fractions of the WD, PEG and DS preparations were equally high (ca. 0.90), whereas both those of the xanthan method and those of the untreated yolk were lower (ca. 0.70). A similar profile was found in investigating the final product.

Though the conclusions drawn from the two investigations differ in detail, the authors agree that the water-dilution method is best.

**Table 7.** Comparison of Efficacy of Several IgY-Isolation Procedures Concerning Yield and Purity

| Test material | IgY | | Test material | IgY | |
|---|---|---|---|---|---|
| (supernatant fluid) | Yield in % | Purity in % | (final product) | Yield mg/ml yolk | Purity % |
| Yolk | 100 | | | | |
| WD-SN | 91 | 31 | WD-UF | 9.8 | 94 |
| PEG-SN | 47 | 25 | PEG-Alc | 4.9 | 89 |
| DS-SN | 71 | x | DS-As2 | 7.5 | 87 |
| Xan-SN | 72 | 32 | Xan-As2 | 7.3 | 89 |

Protein determination by the Biuret method; quantification of IgY by RID (radial immune diffusion, Mancini-technique); check of purity with SDS-PAGE

Special attention is to be given to the pH-value of the distilled water, since Akita and Nakai (1992) found that with a neutral pH the yield of IgY lies between 55 and 60%, whereas with a pH of 5.0 - 5.2 it lies between 92 and 96%.

Moreover the authors of both investigations agree on the simplicity of the water-dilution method in terms of time, materials and manipulations. As regards the production of protective antibodies, Akita and Nakai (1993) also emphasise the optimal opportunities for scaling up.

### Conservation and stability of IgY extracts

Of importance for the use of yolk IgY is information about resistance to heat, for instance in spray-drying for obtaining therapeutic preparations (Ikemori et al., 1997), and about resistance to acidic pH (see 'affinity chromatography').

As regards resistance to heat, Lösch et al. (1986) found no loss of antibodies in eggs cooked for 6 minutes.

On adding HCl in portions, Schmidt et al. (1989) found a rapid fall in the amount of antibodies from pH 4 on. This inactivation of antibody activity could not be confirmed by using glycin/HCl pH 2.3-2.5 to split bound antibodies in affinity chromatography (Schwarzkopf, 1994).

IgY-antibodies can be conserved for about as long as mammalian IgG-antibodies. Various groups of researchers have reported that aliquoted antibodies have been used over several years without any notable loss of activity. The successful use of 10 year old IgY-antibodies has been reported. Systematic investigations have been carried out by Sasse (1998) for instance, who observed the activity of an IgY-antibody in an ELISA over a period of 12 months. The antibodies were aliquoted under sterile conditions and kept at various temperatures (+4°C, -20°C, -70°C). (No sodium azide or thimerosal was added to the aliquots to save them from bacterial infection). Under these conditions the antibodies showed no loss of activity in the course of one year with the exception of -70°C (see Figure 9). Even repeated freezing and thawing at -20°C did not affect the activity of an IgY-antibody in an ELISA (see Figure 10) Antibodies kept at -70°C showed a great loss of activity after a year, the reason for that is not clear.

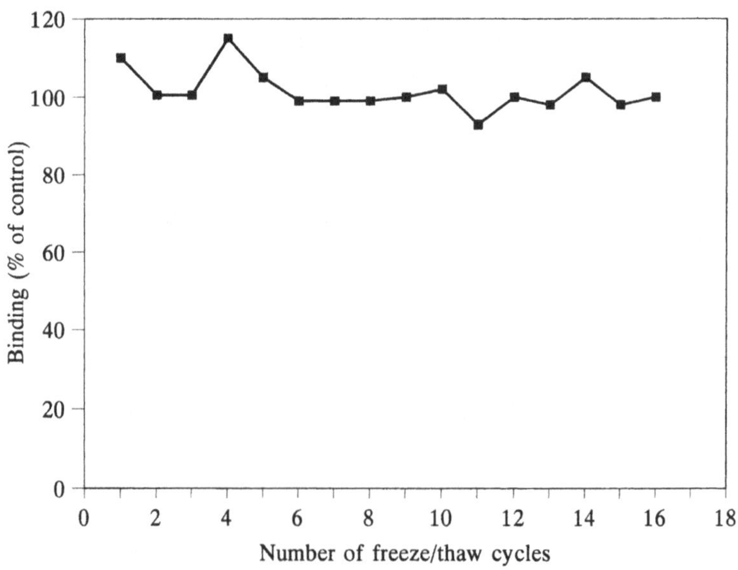


**Fig. 9.** The influence of repeatedly thawing out and freezing an IgY-antibody on its activity in EIA. The comparison is between the activity (extinction) of an antibody after being thawed out once with its activity after being thawed out repeatedly. The test-system was an EIA for CRP-determination (see Chapter 5). Values were measured over a period of about 30 days.

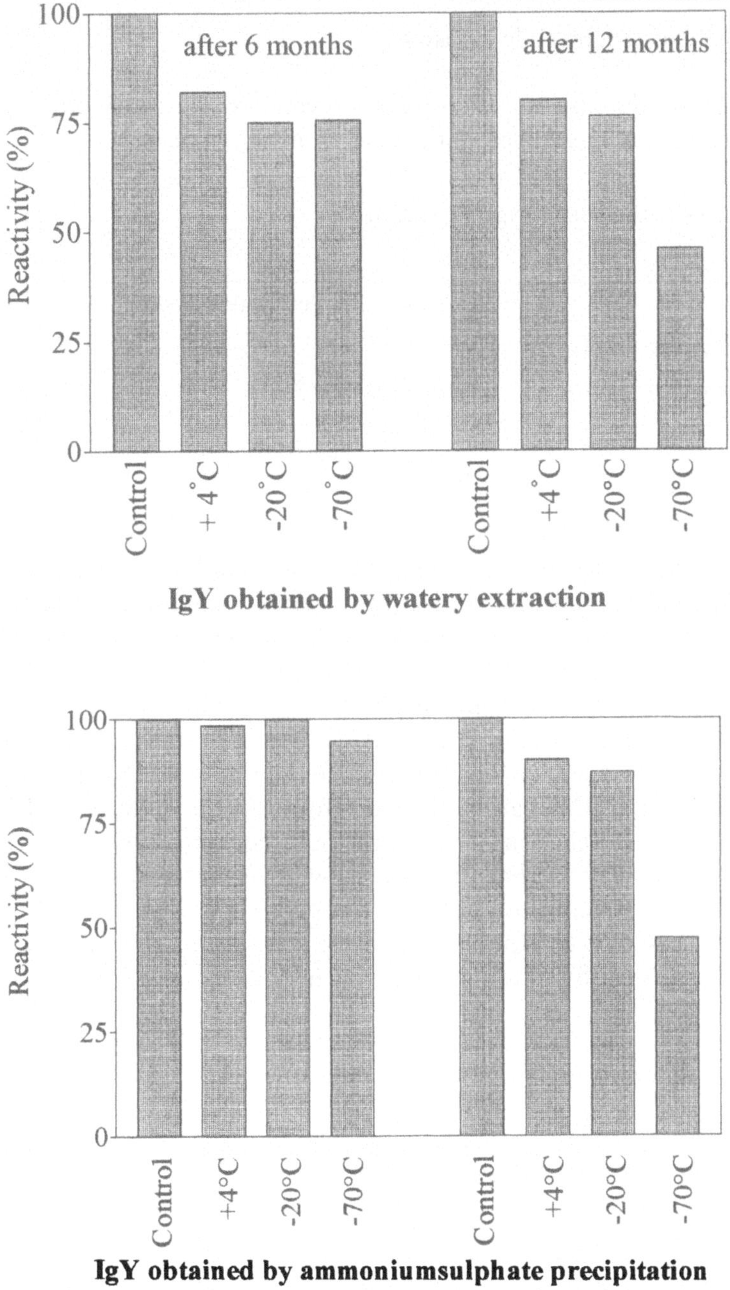


**Fig. 10.** The activity of an IgY-antibody kept at various temperatures. For the upper part of the figure use was made of a watery IgY-extract, whereas for the lower part use was made of an ammonium-sulphate preparation. The test system used was an EIA for detecting the lipopolysaccharide of E. coli 0111 (homologous antigen).

### Systems commercially available

Unlike the case several years ago, several firms are now offering extraction kits for preparing IgY from egg-yolk.

**Table 8.** Commercial IgY Extraction Kits

| Firm | Kit |
|---|---|
| Promega | EGGstract© IgY Purification System |
| Pharmacia Biotech | gammaYolk |
| Clontech | Protein L agarose beads |
| Biotez | IMSORB Anti IgY |
| Technogen | Kaptiv-GY TM |

The first two kits represent precipitation procedures, whereas the third kit works like the affinity-chromatographical purification of IgG with protein G or A. Here too the material involved is a bacterial cell-wall protein from *Peptostreptococcus magnus*, protein L, which is able to bind the immunoglobulins of vertebrates. Unlike protein A/G, protein L does not interact with the Fc-part of the immunoglobulin molecule but with the light chain (Chateau et al. 1993. However, the authors of this paper themselves note that protein L does not bind chicken IgG. We have tested this system with no success, it may be that the system does not work properly). IMSORB anti IgY is also based on an affinity-chromatographic principle. IgY is simply purified by binding to an anti IgY Ab.

## References

Akita EM, Nakai S (1992) Immunoglobulin from egg yolk: Isolation and purification. J Food Sci 57:626-634

Akita EM, Nakai S (1993) Comparison of four purification methods for the production of immunoglobulins from eggs laid by hens immunized with enterotoxigenic E.coli strain. J Immunol Methods 160:207-214

Aulisio CG, Shelokov A (1967) Substitution of egg yolk for serum in indirect fluorescence assay for Roux Sarcoma virus antibody. Proc Soc Exp Biol Med 126:312-315

Bade H, Stegemann H (1984) Rapid method of extraction of antibodies from egg yolk. J Immunol Methods 72:421-426

Chateau, M de, Nilson, BHK, Erntell, M, Myhre, E, Magnusson, CGM, Akerström, B, Björck, L (1993). On the interaction between protein L and immunoglobulins of various mammalian species. Scand. J. Immunol. 37, 399-405
Czerny CP, Eichhorn W (1989) Characterization of monoclonal and polyclonal antibodies to bovine enteric coronavirus: establishment of an efficient ELISA for antigen detection in feces. Vet Microbiol 20:111-122
De Leuw (1996) Gewinnung und Einsatz multispezifischer Antikörper als Hilfsmittel zur Probenvorbereitung in der Rückstandsanalytik von Penicillinen. Dr. Thesis, Bergische Universität-Gesamthochschule, Wuppertal
Erhard MH, Bergmann J, Renner M, Hofmann A, Heinritzi K (1996) Prophylactic effect of specific egg yolk antibodies in diarrhea of weaned piglets caused by Escherichia coli K88 (F4). J Vet Med A 43:217-223
Fischer M (1996) Untersuchungen zu Standardpräparationen von Immunglobulinen aus dem Dotter von Eiern verschiedener Wirtschaftsgeflügelspezies. Dr. Thesis, Veterinärmedizin der FU Berlin, Journal Nr. 1988
Fischer M, Hlinak A, Montag T, Claros M, Schade R, Ebner D (1996) Vergleich von Standardmethoden zur Präparation von Dotterantikörpern. Tierärztl.Praxis 24:411-418
Gassmann M, Thömmes P, Weiser T, Hübscher A (1990) Efficient production of chicken egg yolk antibodies against conserved mammalian protein. Faseb J 4:2528-2532
Gast RK, Porter RE, Holt PS (1997) Assesing the sensitivity of egg yolk antibody testing for detecting Salmonella enteritidis infections in laying hens. Poultry Sci 76:798-801
Gervelmayer A (1995) Herstellung von virusspezifischen vitellinen Antikörpern und deren Einsatz im indirekten Fluoreszenztest zum Nachweis der Erreger der Geflügelkrankheiten Newcastle Disease, Infektiöse Bronchitis und Gumboro Disease. Dr. Thesis, FU Berlin, Journal Nr. 1835
Gottstein B, Hemmler E (1985) Egg yolk immunoglobulin Y as an alternative antibody in the serology of echinococcosis. Z Parasitenkd 71:273-276
Gross M, Speck J (1996) Aviäre Dotter-Antikörper in Diagnostik und Forschung. Dtsch tierärztl Wschr 103:417-422
Grossfeld (1938) "Handbuch der Eierkunde" Verlag v. J. Springer, Berlin
Guiterrez JA, Guirriero V (1991) J Immunol.Methods 143:81-888
Hassl A, Aspöck H, Flamm H (1987) Comparative studies on the purity and specificity of yolk immunoglobulin Y isolated from eggs laid by hens immunized with toxoplasma gondii antigen. Zbl Bakt Hyg A 267:247-253
Hassl A, Aspöck H (1988) Purification of egg yolk immunoglobulins. J Immunol Meth 110:225-228
Hatta H, Kim M, Yamamoto T (1990) A novel isolation method for hen egg yolk antibody, "IgY". Agric Biol Chem 54:2531-2535
Higgins DA, Cromie RL, Liu SS, Magor KE, Warr GW (1995) Purification of duck immuno-globulins: an evaluation of protein A and protein G affinity chromatography. Vet Immunol Immun Path 44:169-180

Holtzhauer M (1997) Biochemische Labormethoden (Springer Labor Manual), Springer Berlin Heidelberg New York

Ikemori Y, Kuroki M, Peralta RC, Yokoyama H, Kodama Y (1992) Protection of neonatal calves against fatal enteric colibacillosis by administration of egg yolk powder from hens immunized with K99-pilliated enterotoxigenic E. coli. Am J Vet Res 53:2005-2008

Ikemori Y, Ohta M, Umeda K, Icatlo FC , Kuroki M, Yokoyama H, Kodama Y (1997) Passive protection of neonatal calves against bovine coronavirus-induced diarrhea by administration of egg yolk or colostrum antibody powder. Vet Microbiol 58:105-111

Jensenius JC, Andersen I, Hau J, Crone M, Koch C (1981) Eggs: Conveniently packaged antibodies, methods for purification of yolk IgG. J Immunol Meth 46:63-68

Kierek-Jaszczuk R, Marquardt R,. Abramson D (1997) Use of a heterologous solid-phase antigen in an indirect competitive antibody-capture anzyme-linked immunosorbent assay for T-2 Mycotoxin. J Food Protect 60:31-327

Kronvall G, Ulysses SS, Finstad J, Williams RC (1970) Phylogenetic insight into evolution of mammalian Fc fragment of gG globulin using staphylococcal protein A. J Immunol 104:140-147

Kuroki M, Ikemori Y, Yokoyama H, Peralta RC, Faustino CI, Kodama Y (1993) Passive protection against bovine rotavirus induced diarrhea in murine model by specific immunoglobulins from chicken egg yolk. Vet Microbiol 37:135-146

Lee SC, Lee KN, Schwartzott DG, Jackson KW, Tae EC, McKee PA (1997) Purification of human alpha$_2$-antiplasmin with chicken IgY specific to ist carboxy-terminal peptide. Prep Biochem Biotechnol 27:227-237

Loeken MR, Roth TF (1983) Analysis of maternal IgG subpopulations which are transported into the chicken oocyte. Immunology 49:21 - 28

Lösch U, Schranner I, Wanke R, Jürgens L (1986) The chicken egg, an antibody source. J Vet Med B 33:609-619

Martin WG, Turner NH, Cook WH (1959) Macromolecular properties of vitellin from egg yolk and its parent complex with lipids. Canad J Biochem Physiol 37:1197-1207

McKinney MM, Parkinson A (1987) A simple, non-chromatographic procedure to purify immuno-globulins from serum and ascites fluid. J Immunol Meth 96:271-278

Mehner A, Hartfield A (1983) Handbuch der Geflügelphysiologie Vol 1, VEB Gustav Fischer Verlag Jena

Mohammed HO, Yamamoto R, Carpenter TE, Ortmayer HB (1985) A statistical model to optimize enzyme-linked immunosorbent assay parameters for detection of *Mycoplasma gallisepticum* and *M. synoviae* antibody in egg yolk. Avian Diseases 30:389-397

Ntakarutimana V, Demedts P, van Sande M, Scharpè S (1992) A simple and economical strategy for downstream processing of specific antibodies to human transferrin from egg yolk. J Immunol Meth 153:133-140

Orlans E, Rose ME (1972) An IgA-like immunoglobulin in the fowl. Immunochemistry 9:833-838

Patterson R, Younger JS, Weigle WO, Dixon FJ (1962) Antibody production and transfer to egg yolk in chickens. J Gen Physiol 45:501-513
Pharmacia, LKB Biotechnology "Affinity Chromatography - principles & methods" Pharmacia LKB Biotechnology AB, Uppsala, Sweden
Polson A (1990) Isolation of IgY from the yolks of eggs by a chloroform polyethylene glycol procedure. Immun Invest 19:253-258
Polson A, von Wechmar MB, van Regenmortel MHV (1980) Isolation of viral IgY antibodies from yolks of immunized hens. Immunol Comm 9:475-493
Polson A, Coetzer T, Kruger J, v.Maltzahn E, van der Merwe KJ (1985) Improvements in the isolation of IgY from the yolks of eggs laid by immunized hens. Immun lnvest 14:323-327
Rammensee M (1984) Immunglobulinkonzentration in Serum, Sekret Ei und Embryo des Haushuhnes. Dr. Thesis, Ludwig-Maximilians Universität München
Romanoff AL, Romanoff J (1949) "The Avian Egg", J Wiley & Sons lNC., New York, Chapman & Hall, Limited, London
Rose ME, Orlan E, Buttress N (1974) Immunoglobulin classes in the hens egg: their segregation in yolk and white. Eur J Immunol 4:521-523
Rosol TJ, Steinmeyer CL; McCauley LK, Merryman JI, Werkmeister JR, Gröne A, Weckmann MT, Swayne DE, Capen CC (1993) Studies on chicken polyclonal anti-peptide antibodies specific for parathyroid hormone-related protein (1-36). Vet lmmunol lmmunopath 35:321-337
Sachsenweber O, Lohr JE, Kösters J (1994) Beurteilung von drei kommerziellen ELISA-Testkits zum Nachweis von Antikörpern gegen *Salmonella enteritidis*. Tierärztl Praxis 22: 350-357
Sasse M (1998). Experimente zur Induktion, Präparation und Charakterisierung aviärer vitelliner Antikörper gegen Endotoxine verschiedener gramnegativer Bakterien. Dr. Thesis, Veterinärmedizinische Fakultät der Universität Leipzig.
Schade R, Pfister C, Halatsch R, Henklein P (1991) Polyclonal IgY antibodies from chicken egg yolk - an alternative to the production of mammalian IgG type antibodies in rabbits. ATLA 19:403-419
Schade R, Schniering A, Hlinak A (1992) Die Gewinnung spezifischer polyklonaler Antikörper aus dem Dotter von Hühnereiern als Alternative zur Immunisierung von Säugern - eine Übersicht. ALTEX 9:43-56
Schwarzkopf C (1994) Gewinnung und immunologische Charakterisierung speziesspezifischer IgY, sowie deren Einsatz zur Bestimmung der Wirtstierart aus dem Abdominalblut hämatophager Insekten. Dr. Thesis, F.U. Berlin
Schmidt P, Wiedemann V, Kühlmann R, Wanke R, Linckh E, Lösch U (1989) Chicken egg antibodies for prophylaxis and therapy of infectious intestinal diseases. II: In vitro studies on gastric and enteric digestion of egg yolk antibodies specific against pathogenic E. coli strains. J Vet Med B 36:619-628
Siewert E, Bronch K (1972) In: Lenkeit, Breirem, Crasemann (eds) Handbuch der Tierernährung , Verlag Paul Parey, Hamburg und Berlin

Staak C, Salchow F, Clausen P.-H, Luge E. (1996) Polystyrene as an affinity chromatography matrix for the purification of antibodies. J Immunol Methods 194:141-146

Svendson, L, Crowley A, Ostergaard LH, Stodulski G, Hau J (1995) Development and comparison of purification strategies for chicken antibodies from egg yolk. Labor Anim Sci 45:89-93

Tjissen P (1985) "Laboratory Techniques in Biochemistry and Molecular Biology: Practice and Theory of Enzyme Immunoassays" Elsevier, Amsterdam, New York, Oxford.

UFAW (1987) The domestic fowl and turkey. In: Poole TB (ed) Handbook on the care and management of laboratory animals. Longman Scientific & Technical. 6th Edition, p.645

Wallmann, J, Staak C, Luge E (1990) Einfache Methode zur Isolierung von Immunoglobulin (Y) aus Eiern immunisierter Hühner. J Vet Med B 37:317-320

Williams J (1963) A comparison of conalbumin and transferrin in the domestic fowl. Biochem H 83:355-364

Yamamoto H, Watanabe H, Sato G (1975) Identification of immunoglobulins in chicken eggs and their antibody activity. Jap J Vet Res 23: 131-40

Yokoyama H, Peralta RC, Horikoshi T, Hiraoka J, Ikemori Y, Kuroki M, Kodama Y (1992) A two step procedure for purification of hen egg yolk immunoglobulin G: utilization Hydroxypropyl-methylcellulose phthalate and synthetic affinity ligand gel (Avid Al). Poult Sci 72:275-281

## Suppliers

**Promega**
Labeled anti-chicken Ab (as secondary Ab) are also available.
Tel.: 0049-(0)621-8501-0
FAX: 0049-(0)621-8501-222
E-mail: de.underscore custserve@de.promega.com
Internet: www.promega.com

**Pharmacia Biotech**
Tel.: 0049-(0)761-4903-153
FAX: 0049-(0)761-4903-353
E-mail: cust.servde@eu.abiotech.com
Internet: http://www.apbiotech.com

**Biotez**
Tel.: 0049-(0)30-9489-3317
E-mail: info@biotez.de
Internet: www.mdc-berlin.de/biotez/de

**Technogen**
Tel.: +39-0823-612-208/228
Fax: +39-0823-612-222/230
E-mail: info@technogen.it
Internet: http://www.technogen.it

# Use of Polyclonal Avian Antibodies

Ingrid Behn, Undine Hommel, Michael Erhard,
Andreas Hlinak, Rüdiger Schade,
Christine Schwarzkopf, and Christian Staak

## Introduction

For many years antibodies have been used for quantitation and detection of a lot of different substances. The specific recognition between antibodies and antigens has led to the development of diverse methods for basic research in the sciences as well as for diagnosis in human and veterinary medicine.

In principle, specific antibodies (immunoglobulins) can be gained by collecting serum from vertebrates immunised with substances (structures) which are antigenic (polyclonal antibodies), or they can be produced by the cell-fusion of antibody-

✉ Ingrid Behn, Universität Leipzig, Fakultät für Biowissenschaften, Pharmazie und Psychologie, Talstr. 33, Leipzig, 04103, Germany (*phone* +49-341-9736-729; *fax* +49-341-9736-749; *e-mail* ibehn@rz.uni-leipzig.de)
Undine Hommel, Universität Leipzig, Fakultät für Biowissenschaften, Pharmazie und Psychologie, Talstr. 33, Leipzig, 04103, Germany
Michael Erhard, Universität Leipzig, Veterinärmedizinische Fakultät, Veterinär-Physiologisches Institut, An den Tierkliniken 7, Leipzig, 04103, Germany
Andreas Hlinak, Universitätsklinikum Charité der Humboldt-Universität, Institut für Pharmakologie und Toxikologie, Dorotheenstr. 94, Berlin, 10117, Germany
Rüdiger Schade, Universitätsklinikum Charité der Humboldt-Universität, Institut für Pharmakologie und Toxikologie, Dorotheenstr. 94, Berlin, 10117, Germany
Christine Schwarzkopf, Zentrales Institut des Sanitätsdienstes der Bundeswehr Koblenz, Laborabt. II (Vet. Med), Generaloberst-Beck-Str. 17, Mainz, 55129, Germany
Christian Staak, Bundesinstitut für gesundheitlichen Verbraucherschutz und Veterinärmedizin, Diedersdorfer Weg 1, Berlin, 12277, Germany

producing cells with permanently growing myeloma cell-lines (monoclonal antibodies).

Traditionally, the species chosen for antibody production have been taken from mammals, especially rabbits, but recently there has been a growing use of hens.

Hen IgY is the main serum antibody, but its transfer to the egg resembles the placental or colostral transfer of IgG in mammals. The active transfer of IgY from serum to eggs leads partly to a higher concentration in the yolk than in the serum. After birds have been immunised, polyclonal antibodies can be prepared in various ways without blood-taking (see Chapter 3 and Chapter 4).

The properties of prepared IgY enable it to be used for manifold specific antigen-antibody reactions, so these antibodies can be used not only for diagnostic purposes but also for therapeutic ones. The immunisation of hens instead of rabbits with antigens from mammals can be advantageous, especially in the case of phylogenetically highly conserved antigens. Prepared IgY-antibodies can easily be marked, for instance with enzymes, fluorochromes and biotin, and both marked and unmarked IgY-preparations can be kept for quite a time under suitable conditions (see Chapter 4 and appendix, respectively).

## Use of polyclonal avian antibodies as primary and secondary antibodies in immunological reactions

This chapter describes several ways of assaying antibodies and antigens. For modern practical purposes, investigators need methods which are specific, sensitive, reproducible, quantitative and quick. Ideally, the methods should need no costly reagents or equipment. They should be effective in measuring many antigens differing in nature and origin, including polyclonal or monoclonal antibodies.

In Subprotocols 1-5 several classical immunoprecipitation techniques are described.

The indirect marking of antigens using an antigen-specific primary Ab and a species specific labeled secondary Ab is advantageous since the wide range of antigen-specific primary Abs (like mouse monoclonal Abs) can be marked by means of only one species-specific labeled secondary Ab. Therefore, methods are included in this chapter which describe in detail procedures to conjugate an IgY Ab with enzymes or fluorescent markers (see Subprotocols 6-8).

In Subprotocols 9-17 several ELISA variants are described in detail, which use the IgY-Ab as labeled or non-labeled primary Ab, as a capture Ab or as a labeled or non-labeled secondary Ab. Simple to perform Dot-Blot methods as well as a Western-Blot procedure based on IgY-Abs are recorded in Subprotocols 18-20.

The suitability of IgY Ab for immunohistochemical methods is demonstrated in Subprotocols 21-24. In Subprotocols 25-27 the advantageous application of labeled IgY-Abs is displayed in FACS-analysis. The Subprotocol 28 exemplaryly describes an IgY-based RIA and in Subprotocol 29 a rather complicated procedure is described, which uses IgY for a complement fixation test, although IgY does normally not activate mammalian complement. Procedures are shown in this subprotocol to circumvent this problem. In Subprotocols 30-32 IgY is used for prophylactic and therapeutic purposes in veterinary medicine. The application of specific chicken egg yolk Abs in this field

**Table 1.** Comparison of the sensitivity of several immunological methods

| Method | Detection limit (mg/l) | Application | |
|---|---|---|---|
| | | Qualitative | Quantitative |
| Precipitation | | | |
| Mancini | 10 – 20 | – | ++ |
| Ouchterlony | 10 – 30 | ++ | – |
| Immunoelectro-phoresis | | | |
| Grabar and Willams | 10 – 20 | ++ | – |
| Rocket | 0.2 – 1.0 | – | ++ |
| Two-dimensional | 0.1 – 1.0 | ++ | + |
| Radioimmunoassay | 0.0001 – 0.01 | – | ++ |
| Enzymimunoassay | 0.001 – 0.1 | – | ++ |

becomes possible because of the high amount of such Abs obtainable from only a few chickens. The use of specific IgY Abs in therapy and prophylaxis of e.g. diseases of the digestive system of mammals in their early life may be an interesting alternative to the use of antibiotics.

A comparison of the detection limits of the various methods is shown in Table 1.

## Subprotocol 1
## Single Radial Immunodiffusion (SRID) - 'Mancini'

Basically, immunoprecipitation and agglutination (e.g. Subprotocols 1-5) are techniques which are simple in nature and require only the participation of antigen and antibody to produce a visible immunocomplex (IC). In agglutination the antigen is of corpuscular nature, in immunoprecipitation it is of non-corpuscular nature. In precipitation, ICs are formed either at the interphase between an antigen containing and an antibody containing fluid or in gel at an area of diffusion where the ratio of antigen and antibody amounts lies within the equivalence range. Due to its growing size, the precipitate can neither diffuse nor migrate electrophoretically in gels, and different precipitation techniques are based on this immobilisation.

Specific IgY from hens differs from mammalian antiserum, for instance, in the fact that maximal immunoprecipitations occur in a medium of 1.5 M NaCl (high salt) (Benedict and Yamaga 1976). Precipitation can be further increased by adding 5% (w/v) PEG 6000 to the medium (Ntakarutimana et al. 1992). The extent to which these manipulations should be used depends on the questions posed. In parallel investigations using 1.5 M and the 'normal' concentration of 0.15 M NaCl, Polson et al. (1980) found no basic difference in the results. Hence in routinely using IgY in immunoprecipitation-techniques, one generally has the advantage of being able to use the rich experiences of classical serology with mammalian antibodies in the physiological medium (0.15 M) without having to introduce a second buffer system.

For immunoprecipitations, gels have proven their worth as semi-solid carriers. Use has been made of both purified agar and agarose. Polson et al. (1980) preferred to use agarose rather than agar, since occasionally white haloes formed round the cup containing the IgY. The authors suspected that these haloes were due to a reaction between agaropectin in the agar on the one hand and traces of lysozym in the not completely purified IgY on the other.

Agar is a mixture of polysaccharides, most of which contain the disaccharide agarobiose. Agarobiose contains many reactive side-groups able to absorb various proteins like fibrinogen, lysozym, thyreoglobulin and ß-lipoproteins. For this reason, agar has been fractionated into agaropectin and agarose. Unlike agar, agarose has no absorbent qualities so it is easier to standardise (Ferencik, 1993).

Antigens are added to wells punched in agarose-gel from where they diffuse radially. They react with antibodies (IgY) distributed evenly in the gel and precipitate round each well as a ring. Within the equivalence range, the width of each ring is proportional to the amount of antigen. Known concentrations of antigens (standard curve) serve as a standard of comparison for the quantification of others.

SRID is used for example for quantifying serum-Ig-isotypes but has other uses too.

SRID is often used when so-called rocket-immunelectrophoresis does not lead to a reliable quantification of the sample, owing to insufficient electropheretic migration of the protein to be determined.

Here is the procedure, as shown by finding the concentration of bovine IgG1 in the serum of cattle:

## Materials

**Equipment**

- gel-punchers, preferably 2 mm in diameter
- a surface to be coated with gel, like
  - glass plates or
  - petri-dishes or
  - microscope-slides or
  - foil-materials

- a temperature controlled waterbath
- a cooker (electrical or gas, for melting the agarose)
- dryer (for drying the gels, e.g. a hair dryer)
- pipettes
- pipette-tips

If it is intended to stain the gel after completion of the test, the carrier glass slides should be coated with melted 0.2% agarose, then baked for 1 h at 100°C in a hot-air steriliser. This is advisable to stop the gel from slipping off the carrier during washing processes.

**Reagents, solutions**

- agarose (e.g. Serva 11397)
- phosphate-buffered saline (PBS, 0.15M), pH 7.2 - 7.4
- precipitating IgY-anti-bovine IgG1. The dilution for use is to be found by trial and error. The example here is 1% in the gel.
- the standard serum from cattle with known IgG1 content. The example here is 10.05 mg/ml.
- polyethylenglycol 6000 (this additive leads to a better precipitate formation)

## Procedure

The Mancini-plates are made as follows:

1. Prepare an agarose solution. For a 1.5% solution, heat 0.75 g agarose in 50 ml PBS to a clear solution. Sodium azide may be added in a concentration of 0.02%.
2. Fill each glass tube with 8 ml, then cool it in a water bath to 56°C.
3. Pre-dilute IgY anti-bovine-IgG1 1:20 (PBS), add 2 ml of the diluted serum to each glass tube and mix well with 8 ml of the fluid gel. The agarose-concentration then comes to 1.2% with an antibody-concentration of 1% in the mixture.

4. Pour the gel onto the levelled microscope-slide to a thickness of about 2-3 mm (10 ml of agarose are enough to coat a surface of 50 $cm^2$ with 2 mm in thickness).
5. (Optional but advisable) Put the plates into a refrigerator for a while, to render the gel firmer.
6. Punch cups of a predetermined pattern in the solidified gel.
7. Fill the cups with cattle-serum, diluted for instance 1:8 or 1:16, as also with a control serum diluted to the same extent in a volume of 2-5μl (dependent on the cup diameter).
8. Incubate the plates in a moist chamber at room temperature for 48 h.

Evaluation is based on the circular precipitation areas. Formation of the precipitate is complete when the latter is enclosed by a white halo.

**Note:** The precipitate can be measured more easily if stained. For staining, the plates are immersed in 1% saline with 0.02% $NaN_3$ for 2 days to remove proteins (with several changes), shortly (1 h) rinsed in distilled water and finally dried. The staining can be done with Coomassie brilliant blue or with other protein-stains like azo-carmine or amido-black according to standard protocols.

**Note:** The period of preparation for staining can be notably shortened.

1. Cover the Mancini-plates with moist filter-paper, leaving no blisters, and with absorbent material on top of it.
2. Place a plate and a suitable weight on the absorbent material for 30 min. The aim is to squeeze water and remaining soluble proteins out of the agarose layer.
3. Remove the pressing device and immerse the plates in tap water for 1 to 2 h. Repeat step 1/2.
4. Dry the plates covered with filter-paper at room temperature or with a hair-dryer.

Evaluation

The reaction area (precipitate) of the sample is compared to the reaction area of a standard as follows:
mg/ml sample = mg/ml control x area of sample / area of control

## Results

One alternative is to take the precipitate's diameter (or diameter squared) as a basis, since this is proportional to the concentration of antigens after full formation of the precipitate which as a rule is reached after 2 days incubation. Under these conditions, a straight line is produced by plotting diameter against concentration (method according to Mancini et al. 1965). Another alternative is to measure the precipitate's diameter already after 6-24 h before full formation (Fahey and McKelvey 1965). In this case the diameter is directly proportional to the logarithm of the concentration of antigens, but the results are less accurate, with a variation coefficient of 10% as compared to about 2% with the Mancini technique. The samples are determined by means of a standard curve. With very good antisera, a detection limit of 1-3 mg/l (of, for instance, IgG or albumin) can be reached (see Table 1).

## Troubleshooting

It is recommended to put the gel plate in a refrigerator for about 20 min. In this way the gel becomes more solidified and may not be so easily damaged by preparing the sample cups. Take care to not inundate the cups with inadequate sample volumes.The precipitate can be evaluated only if the agar-solution has been mixed thoroughly with the antibody-solution otherwise irregularly shaped precipitates may occur (see Figure 1). The temperature of the liquid gel before pouring should stay below 58°C, so as not to destroy antibodies. According to our experience, the precipitate evolves better at room temperature than in the fridge, as with other techniques. For ring-precipitation, the plates must be perfectly horizontal; and the punched wells must be far enough from the edge of the gel to avoid edge-phenomena (see Figure1). For further details see Behn 1991.

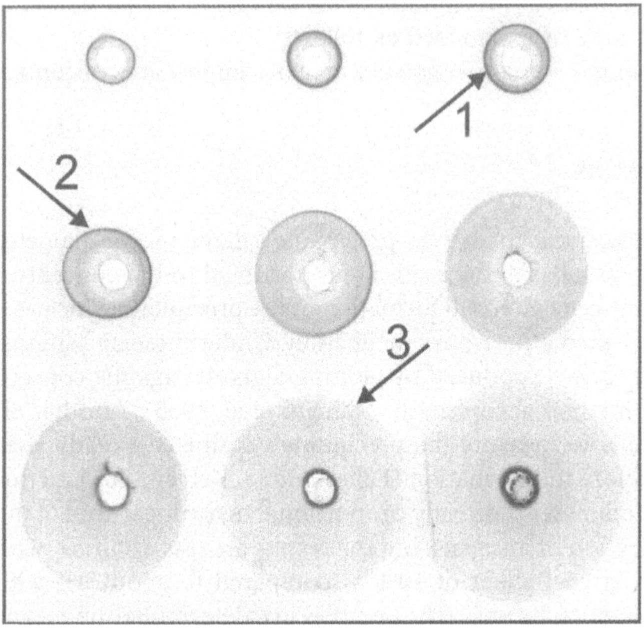


**Fig. 1.** The quantitative determination of various dilutions (1:2-1:512) of bovine IgG1 by means of radial immuno-precipitation (Mancini-technique). The material used was an IgY anti-bovine IgG1 antibody (1.0%). The developed precipitate was stained with Coomassie brilliant blue G250. 1) The clearly marked ring reveals complete precipitation. 2) Lack of a clear ring reveals irregular precipitation and implies a non-horizontal basis. At low concentrations, this effect is mostly negligible. 3) Here there is incomplete precipitation.

## Subprotocol 2
## Agar Gel Double Immunodiffusion Test (AGIDT) – 'Ouchterlony Technique'

In AGIDT, antibodies and antigens from neighbouring cups diffuse radially into the gel and in positive cases create line-shaped immune-complexes. The procedure is suitable for showing the presence of either antibodies or antigens. Antibodies are often titred out, but the results are no more than suggestive. The AGIDT's popularity is due to its use in revealing the presence of infections, and the OIE (Office International des Epizooties, 1996) qualifies it in many cases as a prescribed or alternative procedure for the qualitative determination of antibodies.

## Materials

**Equipment**

- Punchers of various kinds. The most widely accepted pattern consists of 6 punch-holes in a circle around a central cup. These holes are normally 4 mm wide and 8 mm apart (other distances are possible). There are many alternative forms.

Otherwise as given under Mancini-technique.

**Reagents, solutions**

As given under Mancini-technique.

- Equally good results can be obtained by using physiological saline (0.85% NaCl) instead of PBS (see the comments about 'high salt' in the introduction).
- precipitating IgY-anti-CRP
- human CRP-standard serum

## Procedure

**Production of the AGIDT-plates**

1. 1% (w/v) agar in PBS is heated until the solution is completely clear.
2. The hot gel is poured on a glass-slide (or a foil-carrier) in a volume to reach a thickness of about 2-3 mm.
3. After cooling, wells are punched according to a chosen pattern.

**Test procedure**

1. Fill the wells with antigen suitably diluted and others with corresponding antibody in a way that antigen and antibody containing wells are in neighbouring positions. For titrating an antibody, a central well can be filled with CRP-antigen and the surrounding wells with serially diluted anti CRP-Ab. The volume of the wells should range between 5-20µl, depending on the size of the cup. In the case of Figure 2 the Ab is filled in the central well while surrounding wells have been charged with antigen dilutions.
2. Incubate the materials in the moist chamber for 48 h.
3. Evaluate the results. One or more lines of precipitation are to be regarded as a positive result. In the titration of an anti-

body, the dilution at which a precipitation still visibly occurs under constant conditions counts as the titre.

## Results

If the available amounts of antibodies and antigens are small microscopic slides may be used as carrier. A stencil of glass-clear plastic, in which a central reservoir surrounded by further reservoirs, has been cut, is placed on a gel layer about 0.3-0.5 mm thick. The reservoirs are filled with the solutions, with for instance antibodies in the centre and antigens diluted to various extents in the surrounding wells. Through a small hole bored in the base of the reservoir there is a connection to the gel. Usually a blunt needle is pushed through the hole to punch a micro-hole in the gel. The evaluation follows as above.

In qualitative investigations into antigen identities between various antigens in neighbouring cups, note not only the precipitation but also the behaviour of lines of precipitation in the mutual field of reaction. They may create the following patterns.

- Total fusion of both lines in the transition area. This shows that the two antigens are the same (see Figure 2).
- Crossing over without fusion. This shows that the two antigens are different.
- The creation of a spore, if the lines fuse and a spore is created by one of the lines. This shows that the antigens are partly the same.

## Troubleshooting

Unfavourable dilution-ratios of antigens and antibodies can yield negative results, so various ratios have to be tried out.

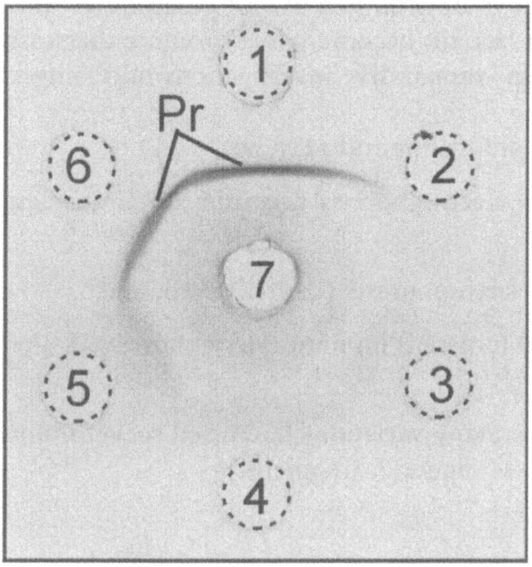


**Fig. 2.** The qualitative determination of rat CRP by means of the Ouchterlony-technique. To the middle hole (7) are added IgY anti-CRP antibodies, and to holes 1-6 are added various CRP preparations in various dilutions (well 1 contains the serum sample, and wells 2-6 contain isolated CRP in various dilutions). Pr = precipitate.

## Subprotocol 3
## Immunoelectrophoresis According to Grabar and Williams (1953)

These methods include both electrophoresis and diffusion (Friemel 1991). They are usually carried out in agarose or agar on glass plates or flexible plastic materials. In the first part antigen substances are separated by means of an electric field, and in the second part mono- or polyspecific antibodies diffuse into the gel and form curved precipitates with homologous antigen fractions or the separated protein fractions migrate electrophoretically into an Ab-containing gel. According to the aim, various kinds of agarose with various qualities (like high purity, low electroendoosmosis and low non-specific absorption) can be used. The agarose can be prepared by means of various buffer-systems (like veronal- or tris-barbital, pH 8-9, ionic strength 0.025, containing 0.02% sodium azide or 0.01% merthiolate as preservatives).

This procedure is of only limited use for quantitative work, and for this reason has not become part of routine diagnosis, but it can be used in comparative investigations into antigens and antibodies.

The procedures still common today are:

- classical immunoelectrophoresis according to Grabar and Williams (1953)
- rocket immunoelectrophoresis (Laurell 1966), and
- two-dimensional (crossed) immunoelectrophoresis (Laurell 1965).

These methods have many variations like fused rocket immunoelectrophoresis (see Figure 7, Chapter 4).

## Materials

**Equipment**

- electrophoresis apparatus (Pharmacia LKB, Multiphor II)
- filter-paper (Schleicher Schüll, No. 334497)
- punch cannule (1-3 mm in diameter)
- carrier (for instance microscope-slide)
- levelled plate
- scalpel (to prepare the reservoires for the Ab)

**Reagents, solutions**

- buffer (for instance a veronal or veronal-sodium buffer, see Chapter 8)
- agarose in electrophoresis buffer (1-2%)
- antigen-dilutions
- Ab-dilutions
- staining-solution (e.g. Coomassie brillant blue, see Chapter 8)

## Procedure

There are two stages, electrophoresis and double diffusion.

**Electrophoresis**

1. Coat the slide with agarose as described.
2. Punch a well in the middle of the carrier and fill it with the antigen to be fractionated electrophoretically.
3. Fill the buffer reservoirs with the appropriate buffer, then use several layers of filter-paper to connect the buffer in the reservoirs with the agarose on the slide.
4. Fractionate the antigen solution by means of an electric current of ca. 5 V/cm for about 60 min.

**Double diffusions**

1. Cut a trench about 2 mm wide along the slide in the agarose on both sides of the punched hole. The pattern may be changed to a central trench with antigen wells to the left and to the right of the trench.
2. Fill the trench with an antiserum suitably diluted, so that the separated antigens and the antiserum diffuse towards each other. This leaves curved precipitates, which may be stained as described above.

## Results

This method is suitable not only for the qualitative demonstration of mixtures of antigens but also for assessing the identities of various proteins (see Figure 3).

## Troubleshooting

The concentrations of the antigen and the antiserum must 'match', and the chosen voltage must not be so great as to warm and shrink the gel. To know when the electrophoresis has ended, mix the antigen solution with bromphenol-blue for instance, as this dye is the quickest to migrate and can serve as a marker.

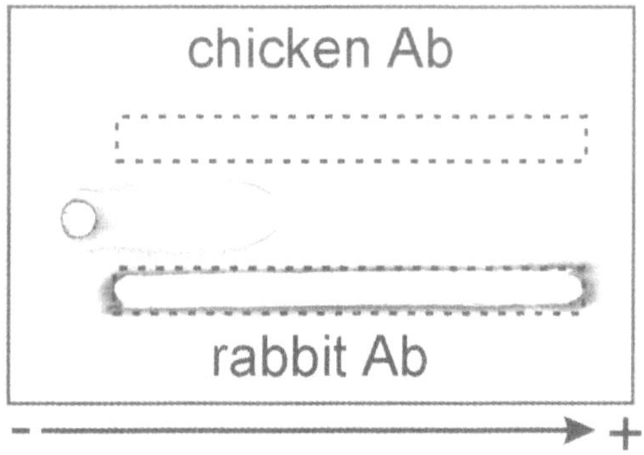


**Fig. 3.** The qualitative determination of rat CRP by means of immuno-electrophoresis. The material applied is a CRP-preparation (antigen), which has been developed against two different antibodies (monospecific anti-CRP antibodies from hens and rabbits). As both precipitate curves join at the end, both antibodies recognize the same antigen (identical specificity).

## Subprotocol 4
## Rocket Immunoelectrophoresis (RIE)

This method, also known as electro-immuno-diffusion, was introduced by Laurell in 1966.

Like simple radial immune diffusion according to Mancini, this procedure is based on the diffusion of antigens into an agarose-gel containing antibodies, but the antigens and antibodies migrate respectively to the anode and cathode in an electric field. To move towards the anode, cathodic migrating proteins (e.g. IgG) must be carbamylated. Rocket-shaped more or less pointed precipitates grow from cups filled with antigens. By including standards of known concentration antigens can be quantified.

### Materials

See Subprotocol 3.

## Procedure

1. Melt agarose to a clear solution, then cool it to about 56°C.
2. Mix the antiserum (which for the quantitative determination of a protein is usually a monospecific antiserum) thoroughly with the agarose. The antibody dilution suitable for this technique must be found by trial and error.
3. Pour the antibody containing gel onto the carrier, cool until the gel has solidified.
4. At a satisfactory distance from the edge of the carrier (about 1 cm, needed for the bridges of filter-paper), punch a row of wells 2 mm wide and 5 mm apart. Provide each plate with sufficient wells for standard samples.
5. Connect the prepared plate with cotton-wicks (filter-paper), as described above, then under slight voltage add the samples to be investigated to the holes.
6. After filling the holes, apply a voltage of 10-15 V/cm. Depending on the electrophoretic properties of the antigen, the rocket develops completely within 30-180 min.
7. Evaluate the plates directly or after staining. The height of the precipitate is directly proportional to the concentration of the antigen and, given the curve of calibrations, can be evaluated simply with the help of an appropriate computer-program.

## Results

The RIE (see Figure 4) can also be used the other way round by putting the antigen in the gel, and the antibodies in the well. In this form, the method is sensitive enough to be excellent for identifying serum proteins or other antigens.

## Troubleshooting

Too high a voltage can heat the agarose and cause it to shrink. Larger sample numbers should be added to the holes under low

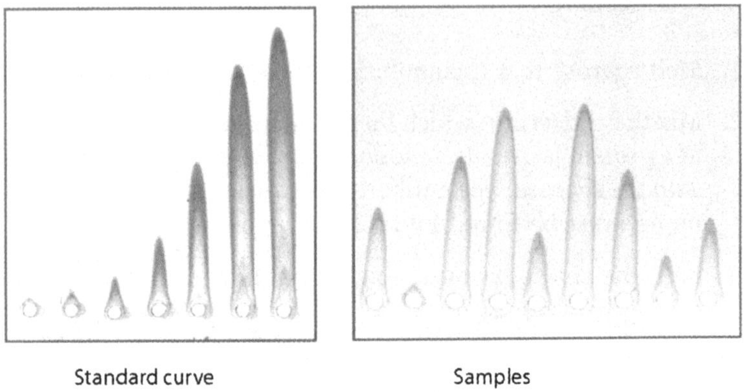


**Fig. 4.** The quantitative determination of rat CRP by means of rocket immunoelectrophoresis. Use was made of a monospecific IgY anti-rat CRP antibody (1%). The left part of the figure shows an example of a standard series (dilutions 1:1024-1:16), and the right part shows the determination of various samples (rat serum 1:10 diluted).

voltage, otherwise the samples added first may start to diffuse radially and to precipitate. They are then not available for forming the rocket-shaped precipitate, so the results are distorted.

## Subprotocol 5
## Two-Dimensional (Crossed) Immunoelectrophoresis

The method is a combination of simple electrophoresis and RIE and was developed in the sixties (Ressler 1960, Laurell 1965, Clarke and Freeman 1966).

### Materials

**Equipment** As Subprotocol 4, further, a knife or other suitable means to cut gel strips (containing electrophoretical separated antigen-mixtures) and to transfer them on to glass plates about 10 x 10 cm. Other sizes and materials may be used.

**Reagents, solutions** As Subprotocol 4.

## Procedure

There are two stages, first dimension-electrophoresis and second dimension-crossed electrophoresis.

**First dimension-electrophoresis**

Carry out a simple electrophoresis on the samples to be investigated in the agarose-gel (see above). Separation is also possible in other gels (PAG) or with other procedures like isoelectric focussing.

**Second dimension-crossed electrophoresis**

1. After completion of electrophoresis, cut off a strip of gel (2 x 10 cm for instance) containing the separated antigens and transfer it to an empty glass plate.
2. Cover the free part of the glass plate with agarose containing a corresponding antiserum.
3. Carry out a second electrophoresis at right angles to the first (i.e. crossed IE). The conditions are the same as for RIE.

According to their electrophoretic mobility, various precipitates form, where the areas or heights at the completion of electrophoresis are proportional to the concentrations of antigens.

## Results

The method is suitable for evaluating mixtures of antigens - like the serum-profile (see Figure 5)- both qualitatively and semiquantitatively under many conditions. It has variants for identifying antigens in a mixture, like tandem two-dimensional IE, intermediary gel two-dimensional IE and two-dimensional line IE.

## Troubleshooting

Pay attention to the relation between the height of voltage and the rise in temperature. The contact between the transferred and the newly poured gel should be as near as possible, with no overlapping. Otherwise as with Subprotocols 1-4.

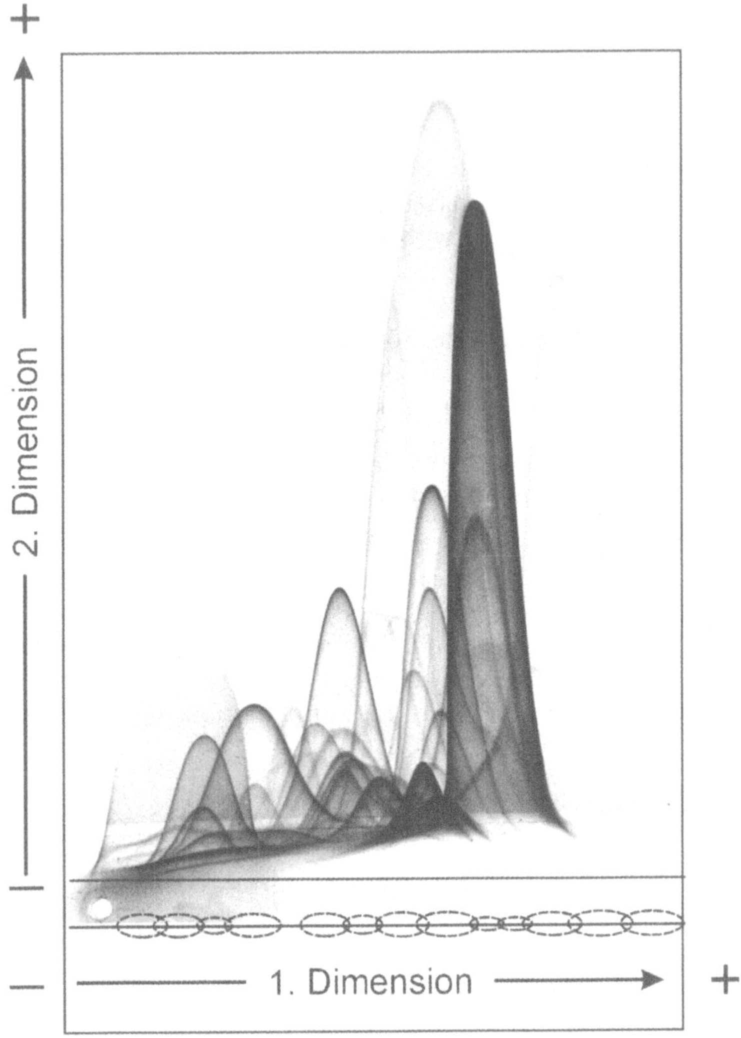


**Fig. 5.** The protein profile of rat serum (diluted 1:4, 1st dimension). A polyspecific IgY anti-rat serum antibody served as the antibody (4%, 2nd dimension).

## Subprotocol 6
## Labeling IgY-Antibodies with Biotin

Biotin is a naturally occurring vitamin with a molecular weight of 244 daltons and an extremely strong affinity for avidin and streptavidin. It can easily be coupled to antibodies via an N-hydroxysuccinimide ester without disturbing the antibodies' properties. (Method modified according to Ey et al., 1978)

### Materials

**Equipment**

- magnetic stirrer
- pipettes and tips
- plastic tubes
- dialysis bags
- shaker

**Reagents, solutions**

- IgY-protein solution (1 mg/ml in PBS or egg buffer, see Chapter 8)
- biotin amidocaproat N-hydroxysuccinimid ester solution (fresh!) (1 mg /100 μl DMSO)
- 3 M glycine
- PBS

### Procedure

IgY-antibodies are mixed with biotin solution and cross-linked by incubation.

**Note:** The biotin solution should be freshly prepared (immediately before the incubation starts).

1. Transfer IgY-protein solution into plastic tube (1 ml)
2. Add biotin solution (25 μl)
3. Incubate 1 h at room temperature while slowly rotating.

4. Stop the reaction and block non-reacted groups for 5 min by adding 80 µl 3 M glycine (rotate slowly).
5. Dialyse the preparation overnight against PBS.
6. Dilute with PBS (1:20) in portions.
7. Store at -20°C.

## Results

IgY-biotin conjugates are tested against a panel of proteins to ensure specificity, reactivity and cross-reactivity. The optimal working-dilution in direct and indirect assays should be determined in a titration.

# Subprotocol 7
# Labeling IgY with Fluorochrome

The conjugation of fluorescein isothiocyanate (FITC) to antibodies is an extremely valuable technique for identifying cell surface molecules by using either fluorescence microscopy or flow cytometry. Fluorescein derivatives are coupled to amino groups of the antibody molecule (IgY). Method modified according to Reisher and Orr (1968, see also Goding 1976, Kronick 1986, and Mohr 1991).

## Materials

**Equipment**

- chromatographic equipment
- photometer
- G25 column (Pharmacia)
- centrifuge

**Reagents, solutions**

- carbonate buffer pH 9.6 (FITC labeling buffer)
- IgY-protein solution (10 mg/ml) dialysed against carbonate buffer pH 9.6 at 4°C with two or three changes

- PBS
- fluorescein isothiocyanate (FITC, Isomer J, adsorbed to Kieselgur, Serva)
- 5 ml Sephadex G25 (Pharmacia)
- dextran blue

## Procedure

1. Mix the dialysed protein solution (100 µl) with 1mg FITC adsorbate.
2. Incubate for 30 min at room temperature while slowly rotating in the dark.
3. Separate the free from bound FITC by means of chromatography on a Sephadex G25 column (with PBS). Dextran blue serves as a marker (due to a mol weight of approx. 2000 kDa dextran blue runs in front of the eluate).
4. Determine and record $E_{280}$ and $E_{495}$.
5. Determine the working dilution.
   Determine the fluorochrome/protein (F/P) ratio. Mol FITC/protein = $(2.87 \times E_{495}) / (E_{280} - 0.35 \times E_{495})$.
6. Store at -20°C.

## Results

IgY-FITC conjugates are tested against a panel of cells to ensure specificity, reactivity and cross-reactivity. F/P ratio should be 3-4 Mol FITC per Mol protein. Other fluorochromes for conjugation are possible, especially rhodamine (TRITC), phycoerythrin and Cy3.

## Troubleshooting

The immunoglobulin fraction should be pure, to avoid possible fluorescence due to fluorochrome's binding to accompanying proteins like albumin.

The optimal working dilution in direct and indirect assays should be determined in a titration.

## Subprotocol 8
## Labeling IgY with Enzyme
## (e.g. Horse Radish Peroxidase - POD)

Whether used qualitatively or quantitatively, enzyme antibody conjugates are most often used via enzyme-substrate reactions which yield a measurable colour. The variety of enzyme and substrate systems enables conjugates to be used in a lot of different enzyme-linked immunosorbent assays as well as in other tests (e.g. in blot systems or immunohistochemistry). Method according to Nakane and Kawoi 1974).

### Materials

**Equipment**

- photometer (Ultraspec)
- pH-meter
- shaker
- stirrer
- pipettes
- pipette-tips
- dialysis-bag
- centrifuge
- refrigerator

**Reagents, solutions**

- IgY-protein solution, dialysed against PBS
- PBS
- distilled water
- horse-radish peroxidase (RZ about 3.0)
- 0.1 M sodium m-perjorate solution

- 0.001 M sodium acetate buffer pH 4.4
- 0.1 M sodium borohydride solution
- 0.2 M sodium carbonate solution
- ammonium sulphate solution, cold-saturated with pH 7.4
- 50% (v/v) ammonium sulphate solution (made from cold-saturated ammonium sulphate solution)

## Procedure

1. Dilute 5 mg POD with 1 ml distilled water and activate POD with 200 ml 0.1 M sodium m-perjorate solution.
2. Incubate and rotate for 20 min at room temperature.
3. Dialyze overnight at 4°C against 0.001 M sodium acetate buffer pH 4.4.
4. Determine the content of POD photometrically after dialysis by $E_{403}$ in Ultraspec.
5. Mix the activated POD and IgY-sample in a molar relation 1:2.
6. Quickly adjust the pH-value to 9.6 by means of 0.2 M sodium carbonate solution.
7. Incubate and rotate for 2 h in the dark at room temperature.
8. Stop the reaction with 100 ml freshly prepared 0.1 M sodium borohydride solution.
9. Incubate at 4°C for 2 h.
10. Add, drop by drop, cold-saturated ammonium sulphate solution (pH 7.4) in the ratio 1:1 (v/v).
11. Incubate overnight in refrigerator.
12. Centrifuge at 4000 g for 10 min.
13. Wash the sediment thrice with 50% saturated ammonium sulphate solution and centrifuge as above.
14. Dilute the enzyme-antibody conjugate with 4 ml PBS.

15. Dialyse intensively against PBS for 24 h.
16. Determine the molar relation of the POD to IgY by measuring $E_{403}$ and $E_{280}$.
17. Dilute the enzyme-protein conjugate with PBS/10% foetal calf serum to a concentration of 0.1 mg POD/ml.
18. Store at -20°C in portions or lyophilise.

## Troubleshooting

Storage of the preparation at -20°C is recommended instead of lyophilisation.

## Comments

Determine the optimal working dilution in direct and indirect assays in a titration. IgY-POD conjugates are tested to ensure specificity and cross-reactivity, and various combinations are suitable as enzymes and substrates for ELISA.

Frequently used marker enzymes and the corresponding substrates

| Enzyme | Substrate |
|---|---|
| (I) horse-radish peroxidase | 2,2'-acino-bis(ethylbenzthiacoline-6-sulphonic acid) (ABTS) 405 nm |
| | o-phenylenediamine (OPD) 492 nm |
| | 3,3',5,5'-tetramethylbenzidine (TMB) 450 nm |
| (II) alkaline phosphatase | p-nitrophenyl phosphate (pNPP) 405 nm |

## Subprotocol 9
## Direct ELISA to Detect Antigens with a Labeled Antigen-Specific IgY-Antibody

An enzyme-immunoassay (EIA) is any immunoassay in which a substance's concentration is found through the activity of a marker-enzyme. In EIA there are basically two reactions: one immunologic and the other enzymic.

There are many assay-variants, classed roughly according to the reaction-principle:

- non-competitive techniques, with direct or indirect assays (sandwich-assays)
- competitive techniques, with the substance competing with a marked tracer for the same binding site

Only competitive assays rely on a single binding-site.

Moreover, according to the sequence of immunologic and enzymatic reactions, a line can be drawn between homogeneous and heterogeneous assays:

- In a homogeneous assay, both reactions take place in a homogeneous solution.
- In a heterogeneous assay, they take place in different solutions (direct and indirect two-sided assays).

Finally the concept EIA applies to the determination of antigens or haptens, whereas ELISA (enzyme-linked immunosorbent assay) applies to the determination of specific antibodies (see, for instance, Engvall 1980, Tijssen 1985, Porstmann and Porstmann 1991). But in the literature, the terms are used loosely at times. In the following we use the term ELISA.

### Outline

Figure 6 shows a schematic view of the assay principle.

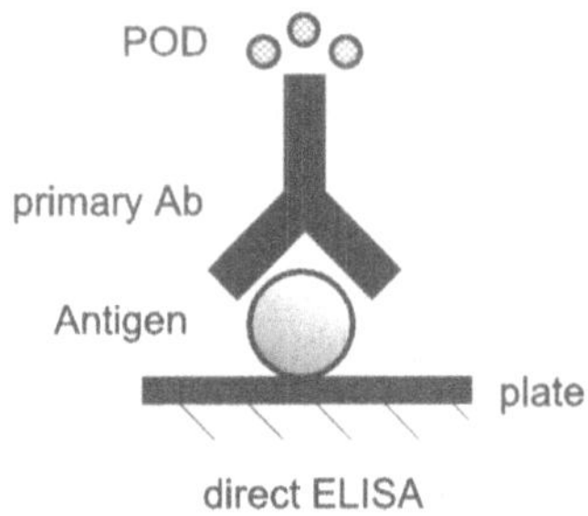


**Fig. 6.** Scheme for a direct ELISA.

## Materials

**Equipment**

- microtitre-plates (from different suppliers, e.g.Flow Laboratories, Linbro No.76-381-04)
- microtitre-plate reader (e.g. Tecan, Spectra)
- microtitre-plate washer (e.g. Corning, ELISA plate washer)
- multi-channel pipettes (e.g. Bibby Dunn, Labortechnik GmbH)
- pipette tips
- incubator

**Reagents, solutions**

- coating buffer (carbonate buffer 0.05 M, pH 9.6 or PBS)
- wash buffer, PBS, pH 7.2-7.4/0.05% Tween 20 (Triton-X 100 can be used instead of Tween 20, different wash buffer)
- stop solution, like sulphuric acid (1M)
- blocking solution (to block non-specific binding, several alternatives are possible e.g. 0.5% gelatine, 10% horse serum, 10% gelafusal [Serumwerke Bernburg], 10% haemaccel [Behring-Werke], 1% BSA)
- substrate solution for POD (especially ABTS or TMB or OPD)

- antigen dissolved in coating buffer (1-10 µg/ml, pH 9.6)
- detecting antibody (IgY-POD-labeled) diluted with PBS pH 7.2/0.05% Tween 20

## Procedure

1. Coat the plate with 100 µl antigen solution (here mouse IgG) to each well.
2. Incubate overnight at room temperature.
3. Wash the plate with 150 µl PBS/0.05% Tween 20 (wash buffer) thrice.
4. Add 50 µl IgY-antibody enzyme conjugate (POD-labeled hen anti-mouse Ig).
5. Incubate for 1h at 37°C.
6. Wash the plate as above.
7. Add 50 µl (ABTS) substrate solution to each well and leave for 15-45 min at room temperature.
8. Read the plates on a microtitre plate reader, using a 405 nm filter.

## Results

This test has been used for determination of the titre of isotype- and subtype-specific antibodies (IgY) against mouse immunoglobulins (Behn et al., 1996).

## Troubleshooting

The choice of microtitre plates is important, as the binding-qualities of plates vary a great deal. Before starting an investigation, try to become fully informed or test various makes. Also take so-called edge-phenomena into account. Rows along the edges (horizontal or vertical) may yield results differing from the rest.

The choice of the substrate must depend on the optical filters available for the measuring instrument. OPD is nowadays less commonly used, owing to the suspicion that it is cancerogenic.

Adjust the colour-reaction's period of development to individual requirements. It can range from a few minutes to more than thirty. (For details see Porstmann and Porstmann 1991).

## Subprotocol 10
## Indirect ELISA to Detect Protein Antigens or Haptens Coupled to a Protein Carrier with an Unlabeled Antigen-Specific IgY-Antibody and a Labeled Secondary Antibody

### Outline

Figure 7 shows a schematic view of the assay principle.

### Materials

See Subprotocol 9.

Additionally, antibody-enzyme conjugate (e.g. POD-labeled rabbit anti-chicken IgY).

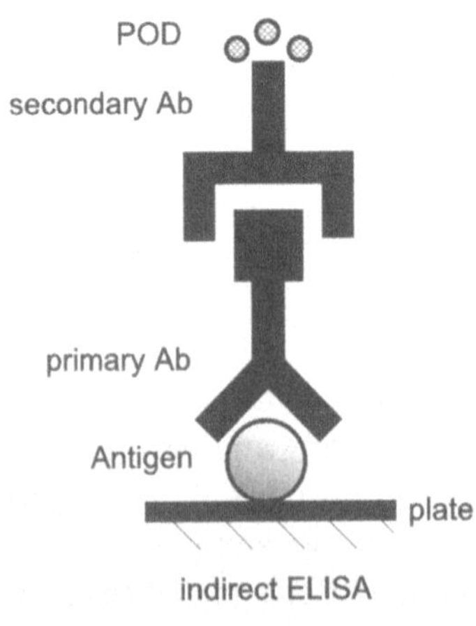


**Fig. 7.** Scheme for an indirect ELISA.

## Procedure

1. Coat the plate with 100 µl antigen in the appropriate dilution (coating buffer) to each well.
2. Incubate overnight at 4°C or for 1 h at 37°C.
3. Wash the plate with 150 µl PBS/0.05% Tween 20 (wash buffer) thrice.
4. Block unspecific binding by incubation with 0.5% gelatine in PBS, 200 µl/well (30 min - 1 h at RT or 37°C).
5. Wash the plate as above.
6. Add 50 µl IgY-antibody preparation (hen anti-x, see Results) suitably diluted to each well and incubate at room temperature or for 1 h at 37°C.
7. Wash the plate as above.
8. Add 50 µl antibody-enzyme conjugate (POD-labeled rabbit anti-chicken IgY) suitably diluted to each well and incubate at room temperature or for 1 h at 37°C.
9. Wash the plate as above.
10. Add 50 µl (ABTS or TMB or OPD) substrate solution to each well.
11. Incubate for 15-45 min at room temperature.
12. Stop the reaction with sulphuric acid (e.g. 1M).
13. Read the plates on the microtitre plate-reader using the corresponding filter (e.g. ABTS-405 nm).

## Results

With this assay, various antigens can be determined qualitatively and quantitatively. In the following, only a few references are given. For further references, see (Chapter 8).

- Detection of various isotypes of mouse immunoglobulin by using hen anti-mouse IgM or IgG (Behn et al. 1996).

- Detection of DNP-haptens on bovine gammaglobulin (BGG) as carrier with hen-anti DNP-human serum albumin (HSA) (Behn et al. 1996).
- Detection of p23 (Hlinak et al. 1996).
- Detection of sCD14 (Hlinak et al. 1996).

### Troubleshooting

See Subprotocol 9.

## Subprotocol 11
## Direct Sandwich ELISA for Detecting C-Reactive Protein

### Outline

In this assay, the capture-antibody is avian, to avoid misleading reactions with so-called rheumatoid-factors (see Figure 8 for assay principle).

### Materials

See above, buffers as described in Subprotocol 9, additionally:

**Reagents, solutions**

- serum samples from patients with inflammatory disease
- monospecific anti-CRP Ab from hen
- monospecific anti-CRP Ab from sheep (POD-labeled)

### Procedure

The procedure is like those already described.

1. Couple the capture-antibody (100 µl antibody-solution/well, purified avian anti-CRP antibody, 2.5 µg/ml) to the plate with carbonate buffer (pH 9.6).

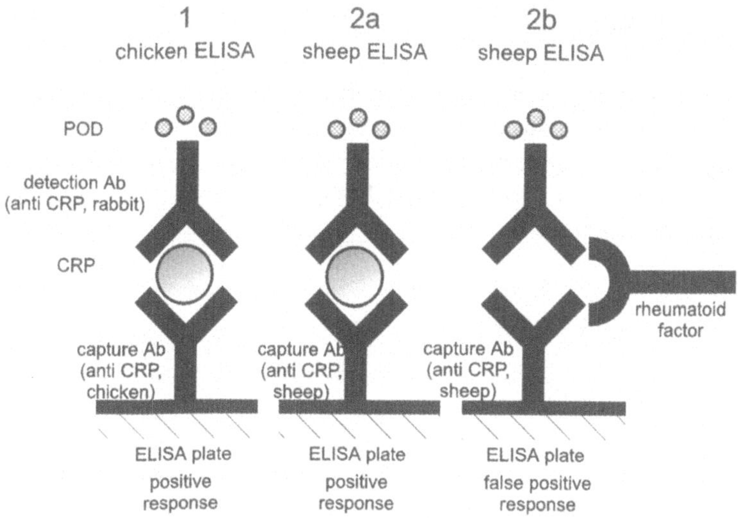


**Fig. 8.** Direct ELISA for determining CRP. There are two versions of the test: *1.* With hen anti-CRP as capture antibody. *2a, 2b.* With sheep anti-CRP as capture antibody. Due to the binding of rheumatoid factors with immunoglobulins there may be false positive reactions with the second version (*2b*).

2. Incubate for 1 h at 37°C (or at room temperature overnight).
3. Wash the plate with PBS-Triton X 100 (pH 7.4, 0.05%, v/v) thrice with 150 μl.
4. Block unspecific binding by incubation with 0.5% gelatine in PBS, 200 μl/well (1 h at 37°C).
5. Add 50 μl of the serum sample (diluted 1:100 with PBS pH 7.4) and incubate for 1 h at room temperature.
6. Wash as above.
7. Add a POD-labeled anti-human CRP antibody (50 μl, Sigma, diluted 1:1.500).
8. Wash as above.
9. Add 50 μl OPD-substrate solution and incubate at room temperature for 15-45 min.
10. Add 50 μl sulphuric acid (e.g. 1M).
11. Read plates on microtitre plate reader using a 492 nm filter.

## Results

The combination of an avian antibody with a mammalian one prevents rheumatoid factors from forming a 'bridge' between capture and primary antibodies (see Figure 8), so results are reliable (Larsson and Sjoquist 1990, Rieger et al. 1996). For details, see Chapter 6.

## Troubleshooting

See Subprotocol 9.

# Subprotocol 12
# Direct Sandwich ELISA for Detecting Bovine IgG2. Use of POD-Labeled Protein G Instead of a Labeled Antibody

This test is suitable for determining the immunoglobulin of mammals. This ELISA functions in this construction only because protein G reacts with immunoglobulin (IgG) of many mammalian species but not with IgY. Protein G also could be used as a binding molecule (instead of a capture antibody) with labeled IgY as the detection-antibody. The assay principle presented here is often less costly than the use of commercial antibodies.

## Outline

Figure 9 shows a schematic view of the assay principle.

## Materials

See Subprotocol 9, additionally:

- anti-immunoglobulin antibodies from chickens
- immunoglobulin in suitable dilution
- protein G (labeled with POD)
- buffers as above

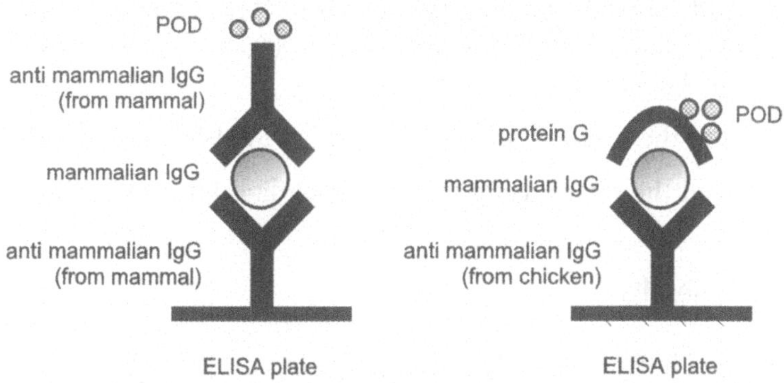


**Fig. 9.** Direct ELISA for determination of mammalian immunoglobulins (IgG). On the left, the capture antibody is a mammalian anti-IgG antibody, and on the right, an anti-IgG antibody from hens. Through this substitution, the bound IgG can be marked with POD-labeled protein G. The latter has no affinity with hen IgY, so false positive reactions are avoided.

## Procedure

1. Coat microtitre plates (Maxisorb, Nunc, Wiesbaden Germany) with affinity-isolated hen anti-bovine IgG2 (100 µl/well, 10 µg Ab/1 ml coating buffer) and incubate overnight at 4°C.
2. Wash the plate with PBS-Tween 20 (pH 7.2, 0.05% Tween 20, v/v) thrice with 150 µl.
3. Block non-specific binding sites with 0.5% gelatine (in PBS, pH 7.2, 200 µl/well), incubate for 1 h at 37°C.
4. Wash as above.
5. Add serum samples from new-born calves or colostrum in suitable dilutions (1:2.000, $log_2$, 50 µl/well) and incubate for 1 h at 37°C.
6. Wash as above.
7. Add POD-labeled protein G (1:2,500 in PBS-Tween, 50 µl/well) and incubate for 1 h at 37°C.
8. Wash as above.

9. Add substrate solution to each well (TMB, 0.2 mg/ml in 0.1 M acetate-citrate buffer, pH 5 with 0.0005% $H_2O_2$, 100 µl/well, see Appendix).
10. Stop the reaction after 10 min with 1 M $H_2SO_4$ (50 µl/well).
11. Read the plates using the corresponding filter (450 nm).

## Results

Purified bovine serum IgG2 (0.5 µg/µl, $\log_2$, 50 µl/well) served as standard.

## Troubleshooting

See Subprotocol 9.

# Subprotocol 13 Sandwich-ELISA With Primary IgY-Antibodies for Detecting *Ascaris suum* - Antigens

## Outline

Figure 10 shows a schematic view of the assay principle.

## Materials

See Subprotocol 9, in addition:

**Reagents, solutions**

- capture-antibody (from rabbit) dissolved in coating buffer (6 µg/ml, pH 9.6).
- extract of samples (aqueous extract of faeces from pigs and other species).
- hen anti-SoAg IgY (somatic antigen of *Ascaris suum*) as primary detection antibody (from hyperimmunised hens) in working dilution.

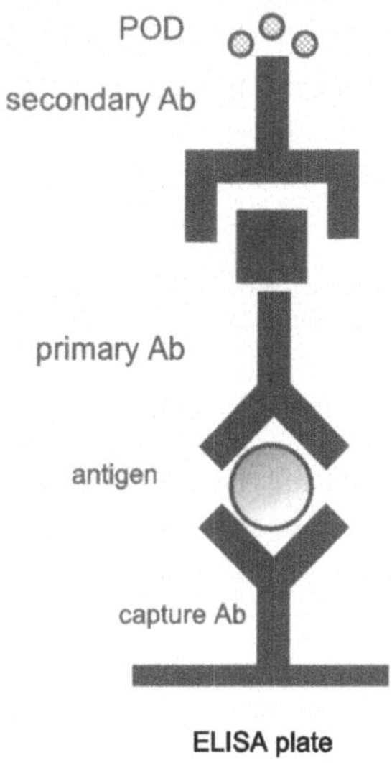


**Fig. 10.** Scheme for an indirect ELISA with a capture antibody as the antigen binding site.

- commercial POD-labeled rabbit anti-chicken IgY in working dilution.

## Procedure

1. Coat the plate with 100 µl capture-Ab solution to each well.
2. Incubate overnight at room temperature.
3. Wash the plate with 150 µl PBS/0.05% Tween 20 (wash buffer) thrice.
4. Block unspecific binding by incubation with 0.5% gelatine in PBS, 200 µl/well (1 h at 37°C).
5. Wash as above.
6. Add 50 µl samples into each well and incubate 1 hr at room temperature.
7. Wash as above.
8. Add 50 µl Ab-preparation in dilution (hen anti SoAg) and incubate 1 hr at room temperature.

9. Wash as above.
10. Add 50 µl POD-labeled rabbit anti-chicken IgY in working dilution.
11. Wash as above.
12. Add 50 µl (OPD) substrate solution to each well and incubate for 15-45 min at room temperature.
13. Stop the reaction with sulphuric acid (e.g. 2.5 N).
14. Read the plates on a microtitre plate reader, using a 492 nm filter.

## Results

This test enables infestation with *Ascaris suum* to be shown clearly (note the time lag according to infestation!) not only in the case of pigs but also of humans. ELISA is more reliable than a coproscopic test (Schniering 1995, Schniering 1996).

## Troubleshooting

For this test it is crucial to add 50% foetal calf serum to the sample, since the proteolytic activity of the faeces-contents may otherwise destroy the layer of capture-antibodies.

# Subprotocol 14 ELISA for Detection of Antibody from Horses Against Dourine (*Trypanosoma equiperdum*)

## Outline

The test system is a typical example for commercial ELISA kits used for control of contagious diseases. The O.I.E. (Office International des Epizooties) has accepted this ELISA as an alternative next to the CFT (complement fixation test). Horse sera are tested in parallel in wells with and without antigen. Differences in extinction between the two are termed "net extinction" (n.Ext.) and a reading ›0.300 is regarded as a positive result.

See Fig. 7 for assay construction.

## Materials

**Equipment**

See Subprotocol 9.

**Reagents, solutions**

- carbonate buffer (see Subprotocol 9)
- phosphate buffer with Tween 20 (PBS, see Subprotocol 9)
- blocking buffer: carbonate buffer plus 3% fetal calf serum (FCS)
- sample and conjugate buffer: PBS-Tween plus 6% FCS
- substrate indicator system: 40 mg ABTS (see Chapter 8) dissolved in 100 ml citric phosphate buffer (see Chapter 8), just before use 100 µl of 1:40 diluted $H_2O_2$ is added to 10 ml of ABTS solution
- field and control sera from horses
- antigen: Trypanosomes are isolated from parasitaemic blood from artificially infected rats according to the method of Lanham and Godfrey (1970). 0.5 ml antigen is reconstituted in 5 ml of carbonate buffer, sonicated 3 x 10 sec at 12 microns peak to peak, and centrifuged at 10,000 x g over 4 min. The supernatant is further diluted in carbonate buffer according to results from pretesting.
- conjugate: POD-labeled IgY anti-horse Ig at a pretested working dilution

## Procedure

1. Charge wells of columns 2, 4, 6, 8, 10, 12 with 50 µl of antigen, and columns 1, 3, 5, 7, 9, 11 with 50 µl of carbonate buffer.
2. Incubate at 37°C for 40 min in a moist chamber.

3. Wash with tap water and afterwards with carbonate buffer.
4. Add 200 µl per well of blocking buffer and incubate at 37°C for 20 min
5. Wash with tap water followed by 4 wash cycles with PBS-Tween.
6. Add sample serum prediluted 1:100 in sample buffer into 2 neighbouring wells (one with antigen and one without antigen), incubate 30 min at 37°C in a moist chamber.
7. Wash as under 5.
8. Add 50 µl of properly diluted conjugate and incubate as under 6.
9. Wash as under 5.
10. Add 100 µl of ABTS/ $H_2O_2$ solution to all wells of the plate. The reaction is stopped by the addition of 25 µl/well of a 1:10 diluted household detergent when the positive standard serum diluted to 1:800 reaches a net-extinction of ›0.300.

## Results

Results are read by a plate reader at 405 nm and results are given as net extinction. The 1:800 diluted standard serum provides a reaction of 100%. Results from the field sera are expressed in percent to the standard reaction.

## Troubleshooting

The system with an integrated control and with the exclusion of non-specific adsorption to the plastic material provides a robust ELISA technique. Misinterpretations can result from reactions caused by contacts with related antigens such like trypanosomes from the T. brucei-group. But these do not occur on the northern hemisphere.

## Subprotocol 15
## Competitive ELISA

ELISA procedures are notably flexible. Most of the methods now used in routine investigations are non-competitive indirect procedures with an immobilised antigen, but in recent years competitive assays have become more common. Some authors draw a line between competitive and blocking ELISA:

In competitive ELISA, serum samples and detection-antibodies are added at the same time to compete on equal terms for specific binding sites.

In blocking ELISA, antibodies present in the sample compete with labeled detection antibodies for the binding sites. If the former predominate, they block all available sites, so enzyme-reactions are weak or absent.

The use of specific IgY (anti-HSP 70) in a competitive ELISA system is illustrated below in the quantitative detection of heat shock protein (HSP 70) (Gutierrez and Guerriero 1991). The idea is that in the pre-incubation of HSP 70 with a standardised concentration of IgY-anti-HSP 70, the latter bind to HSP in proportion to their concentration, so the results in the following ELISA depend on the amount of bound or non-bound anti-HSP 70 IgY (see Figure 11).

### Outline

Figure 11 shows a schematic view of the assay principle.

### Materials

See Subprotocol 9, additionally:

- HSP 70 in suitable dilutions
- chicken anti-HSP 70
- buffers as above

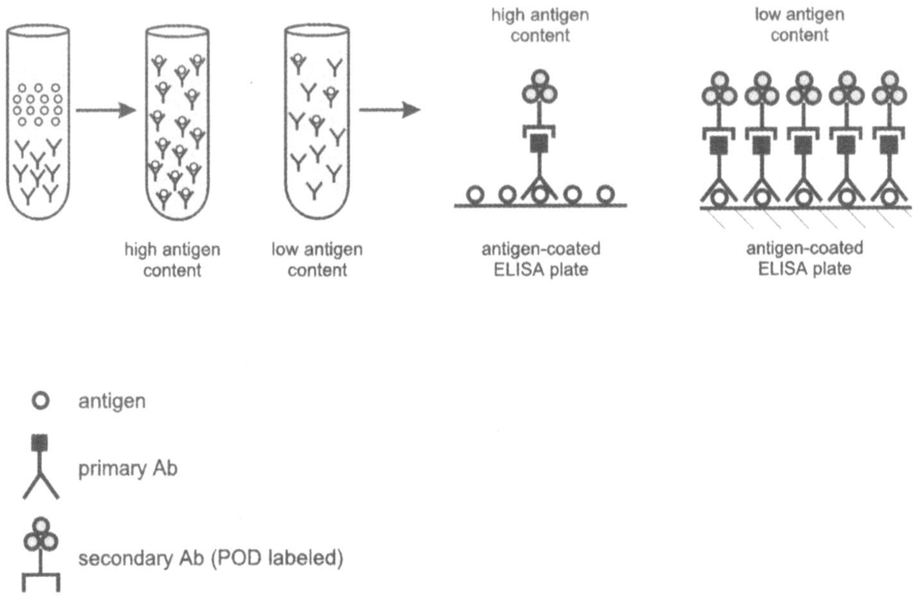


**Fig. 11.** Scheme for a competitive ELISA. What is shown tallies with stage II (preincubation) and stage III (sample transfer) in Subprotocol 15.

## Procedure

**ELISA plates**

Preparation of the ELISA plates:

1. Coat the plate with HSP 70 as the antigen (100 µl antigen solution in coupling buffer).
2. Wash the plate thrice with PBS Tween 20, 0.05%, pH 7.2)
3. Block the plate with gelatine (Tween 20-PBS, 0.1% gelatine).
4. Wash as above.

**HSP 70-standard and samples**

Preparation and preincubation of the HSP 70- standard and the samples:

1. Prepare a HSP 70 stock-solution in a concentration of 800 ng/100 µl.
2. Prepare standard-dilutions as follows: ng/100 µl = 1:400, 1:200, 1:100, 1:50, 1:25, 1:12.5.

3. Add equal volumes of the pre-tested anti-HSP IgY and incubate the mixtures.

Transfer of the incubated samples onto the coated ELISA-plates: **Transfer**

1. Transfer the 6 variously diluted standard and assay-samples, pre-incubated anti-HSP IgY, onto the ELISA plate and incubate at 37°C for 2 h.
2. Wash as above.
3. Add AP(alkaline phosphatase)-labeled rabbit anti-IgY conjugate in working dilution, then incubate them at room temperature for 1 h.
4. Wash as above.
5. Add substrate (p-nitrophenyl phosphate, pNPP) (see Chapter 8).
6. Determine the net-extinction (405 nm) through photometry, then determine the HSP-concentration of the sample by comparing it with the standard set.

## Results

See Figure 11.

In the sample there was a lot of HSP: anti-HSP IgY was wholly bound at stage II, so was unable to react at stage III.

In the sample there was little or no HSP: anti-HSP IgY was hardly or not bound at stage II, so was able to react at stage III.

Competitive ELISA can be used on samples to detect antigens as well as antibodies.

## Troubleshooting

As with other ELISAs.

## Subprotocol 16
## ELISA to Determine the Content of Mouse Monoclonal Antibodies by Means of a Mouse Ig-Specific Capture-Antibody and a Labeled Detecting Antibody from Chicken (Sandwich)

### Outline

For principle see Figure 8.

### Materials

See Subprotocol 9, in addition:

- commercial rabbit anti-mouse Ig (capture antibody solution 1-10 mg/ml PBS)
- monoclonal antibodies of various isotypes (the mAb in this system are the antigens to be detected)
- chicken antibody-enzyme conjugate

### Procedure

1. Coat the plate with 100 µl capture Ab solution into each well and incubate overnight at room temperature.
2. Wash the plate with 150 µl PBS/0.05% Tween 20 (wash buffer) thrice.
3. Add 50 µl solution containing mouse monoclonal antibodies to each well and incubate 1 hr at room temperature.
4. Wash as above.
5. Add 50 µl POD-labeled chicken anti-mouse IgY in appropriate dilution to each well and incubate at room temperature.
6. Wash as above.
7. Add 50 µl ABTS substrate solution to each well and incubate for 15-45 min at room temperature.

8. Read the plates on the microtitre plate reader using a 405 nm filter.

## Results

This assay was successfully used for estimation of content of IgM or IgG in hybridoma cell culture supernatants.

## Troubleshooting

See Subprotocol 9.

# Subprotocol 17
# Indirect ELISA to Detect Antigens With an Unlabeled Primary Antibody and Labeled Secondary Antibody (IgY)

## Outline

See Figure 7.

## Materials

See Subprotocol 9, in addition:

- antigen solution (1-10 mg/ml PBS)
- unlabeled mouse monoclonal antibody in appropriate dilution.
- POD-labeled chicken anti-mouse immunoglobulin in appropriate dilution.

## Procedure

1. Coat the plate with 100 µl antigen solution to be identified to each well and incubate overnight at room temperature.

2. Wash the plate with 150 µl PBS/0.05% Tween 20 as a wash buffer thrice.
3. Add 50 µl unlabeled mAb of different supernatants of hybridoma cell cultures to the wells and incubate for 1 h at room temperature.
4. Wash as above.
5. Add 50 µl POD-labeled chicken anti-mouse immunoglobulin Ab to each well and incubate for 1 h at room temperature.
6. Wash as above.
7. Add 50 µl ABTS substrate solution to each well.
8. Read the plate using the corresponding filter (405 nm).

## Troubleshooting

See Subprotocol 9.

## Applications

This procedure was successfully used for estimation of antibody production of hybridoma cell cultures (screening).

# Subprotocol 18
# Proof of Specificity of Avian Antibodies Used Against CCK (Dot Blot)

The blot assay is a simple way of testing the specificity of antibodies. In this protocol various CCK-derivatives are transferred to foil and developed with anti-CCK Abs of different specificity.

## Outline

See Figure 12.

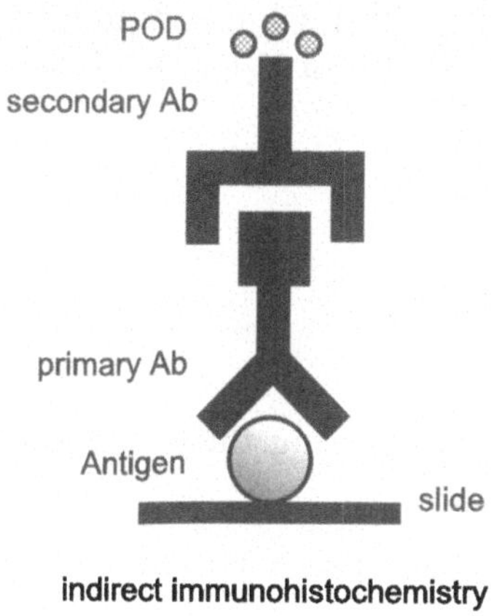


Fig. 12. Scheme for an indirect immunohistochemical marking of an antigen.

## Materials

**Equipment**

- shaker
- pipettes
- pipette-tips
- incubator
- plastic dish

**Reagents, solutions**

- nitrocellulose sheet like NC 2, pore-size 0.2 mm, Serva
- blot buffer (see Chapter 8)
- Triton X 100
- antigen solution (several CCK-derivatives, see Figure 13)
- paraformaldehyde
- skimmed milk-PBS solution (Glücksklee, skimmed milk powder, Nestlé)
- primary IgY-antibody (anti-CCK Ab from hen) in an appropriate dilution (blot buffer)
- secondary anti-IgY antibody, POD-labeled and diluted in a blot buffer

- 15 mg DAB (Sigma) in 50 ml PBS with 1500 ml ammonium nickel sulphate (3% w/v) and with 15 ml $H_2O_2$ (1 ml 30% $H_2O_2$ in 30 ml PBS)
- distilled water

## Procedure

1. Prepare the nitrocellulose strips, handling them carefully (don't touch them).
2. Wash the sheet in a blot buffer (PBS-Triton X 100) for 10 min and shake slowly.
3. Dry the sheet for about 15 min at 37°C in an incubator.
4. Apply the antigen solution in a dilution range, with 1.5 µl solution per dot, to the sheet through a suitable pipette tip.
5. Dry the sheet for 10 min at 37°C.
6. Keep the sheet for 1 h in a paraformaldehyde atmosphere at 80°C in case peptides are detected.
7. Repeat step 2.
8. Block the sheet in a 5% (w/v) skimmed milk PBS solution and shake slowly (1 h).
9. Repeat step 2.
10. Incubate the sheet in the primary IgY-antibody for 1 h.
11. Repeat step 2.
12. Incubate the sheet in the secondary anti-IgY Ab, labeled with POD, for 30 min.
13. Repeat step 2.
14. Stain the sheet for about 10 min or less in a DAB solution.
15. Stop the reaction by means of distilled water.
16. Dry the sheet.

## Results

This is a quick and simple test of the binding of an antibody with a lot of different antigens. Even with protein-antigens, steaming with paraformaldehyde can improve the results (see Figure 13).

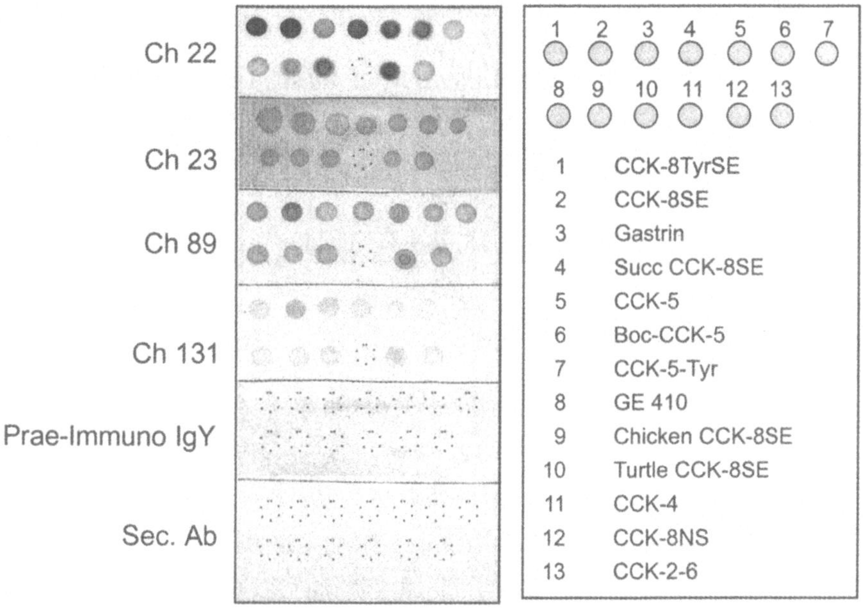


| | |
|---|---|
| **Ch 22** | Asp-Tyr(SE)-Met-***Gly-Trp-Met-Asp-Phe*** -$NH_2$-$(Lsy)_4$-PEG-Resine |
| **Ch 23** | KLH-$(Gly)_4$-Asp-Tyr(SE)-Met-***Gly-Trp-Met-Asp-Phe*** - $NH_2$ |
| **Ch 89** | BSA-Suc-Asp-Tyr(SE)-Met-***Gly-Trp-Met-Asp-Phe*** -$NH_2$ |
| **Ch 131** | BSA-Suc-Asp-Tyr(SE)-Met-***Gly-Trp-Met-Asp-Phe*** -Tyr(SE)-$NH_2$ |

**Fig. 13.** Comparison of the specificity of four chicken antibodies, raised against four CCK-derivatives; specificity testings of various IgY antibodies by means of dot-blot methods (indirect detection). The lower list gives information about the structure of the antigens used for immunisation. The antibodies were used in dilutions of 1:200 - 1:400. The various CCK-derivates (equimolar concentrations) were applied in amounts of 1.5 µl. Dependent on the specificity of the Abs used the various derivatives are marked with various intensities, and antibodies Ch22 and Ch89 distinguish between CCK and gastrin. (CCK and gastrin belong to a peptide family, and have the last 5 amino acids [in italic] in common). SE = sulphated, NS = not sulphated.

### Troubleshooting

Handle the nitrocellulose sheet carefully, without touching it. The sheet must be wholly dry, or the dots will be irregularly shaped. A critical step is the blocking of non-specific binding. We have good experience with low-fat milk powder.

## Subprotocol 19
## Dot Blotting, Detection of Rubella-Virus Proteins

### Outline

See Subprotocol 18 and Figure 12.

### Materials

**Equipment** See Subprotocol 18, additionally:

- scanner

**Reagents, solutions**

- rubella-antigen solution (110 virus units/ml)
- nitrocellulose (BIOMETRA) GM 045
- PROMEGA-Kit A (special buffer concentrate, concentrate: distilled water, 1:1)
- wash buffer I (PBS, 3% skimmed milk powder, 0.2% Tween 20)
- wash buffer II (wash buffer I:solution A, 4:1)
- wash buffer III (solution A:wash buffer I [without Tween 20], 1:1
- substrate buffer (50 ml III + 10 ng DAB + 8 ml $H_2O_2$, 30%)
- 1 M sodium acetate buffer
- monoclonal anti-rubella antibody (primary mouse antibody B1, IgG2a, diluted in wash buffer I)
- POD-labeled detection antibody (secondary chicken anti-mouse IgG antibody) in appropriate dilution (wash buffer I)

## Procedure

1. Prepare the nitrocellulose sheet, handling it carefully (see Subprotocol 18).
2. Apply the antigen solution in a working dilution, with for instance 1 μl solution per dot, to the sheet by means of a pipette with a suitably sized tip.
3. Dry the sheet for 10 min.
4. Wash the sheet for 20 min in blot buffer I and shake it.
5. Incubate the sheet in the primary antibody for 18 h (1:500).
6. Wash the sheet four times in wash buffer I for 15 min.
7. Incubate the sheet in the POD-labeled secondary antibody for 1 h.
8. Wash the sheet four times in wash buffer I.
9. Wash the sheet thrice in wash buffer II.
10. Incubate the sheet once in wash buffer III.
11. Incubate the sheet once in substrate buffer for 15 min.
12. Stop the reaction with 1M sodium acetate buffer for 5 min.
13. Dry the sheet.
14. Evaluate the results by means of a scanner.

## Results

See Figure 14.

## Troubleshooting

See Subprotocol 18.

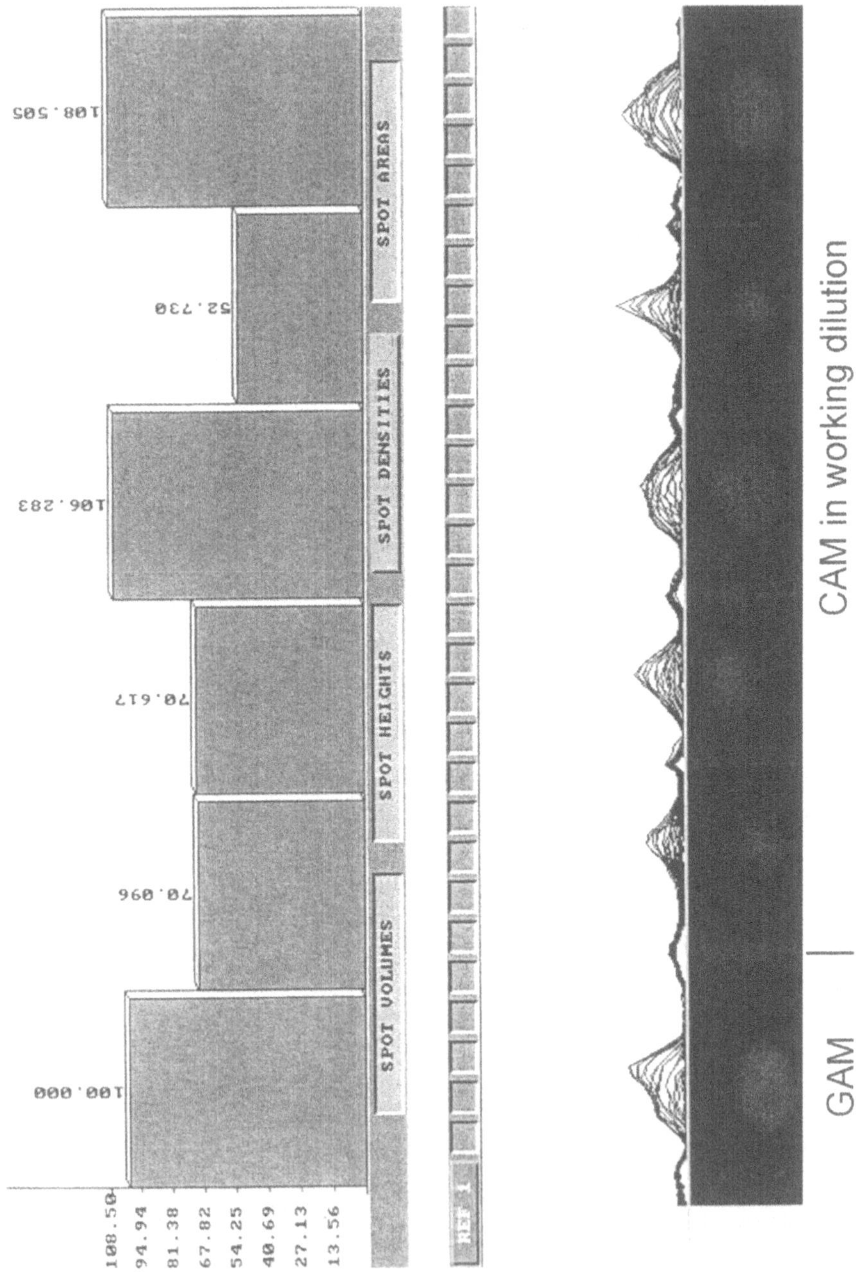


**Fig. 14.** Dot blot method. Demonstration of Rubella virus protein by means of mouse monoclonal antibody. Shown is a comparison between the efficacy of a commercial POD-labeled goat anti-mouse IgG Ab (GAM) and different POD-labeled chicken anti-mouse IgG Ab (CAM). The lower part of the Fig. shows the original spots and the corresponding scans which are expressed in the upper part as a relative value.

## Subprotocol 20
## Immunoblotting, Western Blotting

This method is used frequently to demonstrate a specific agent out of a mixture of several agents or to demonstrate the specificity of an antibody.

### Outline

The principle of the method is based on an electrophoretical separation of a fluid followed by an electrophoretical transfer of the separated fractions on a suited foil. This foil is further treated as described in Subprotocol 18 (dot-blot).

### Materials

**Equipment**

- polyvinylide-difluoride membrane (Immobilon P, Millipore, pore size 0.45 μm)
- electrophoresis chamber (Protean II xi, BIORAD)
- blotting chamber (Trans-Blot-Cell, BIORAD)
- shaker

**Reagents, solutions**

- acrylamide/bisacrylamide stock solution, monomer concentration 30%, cross-link concentration 1% (SERVA, Heidelberg)
- buffer stock-solution (Tris 77.43 g, sodium acetate 43.83 g, EDTA 9.37 g, in 1000 ml dist. water)
- electrophoresis buffer (buffer stock-solution 75 ml, SDS 1.2 g, in 1125 ml dist. water)
- sample buffer (32.5 ml dist. water, buffer stock-solution 3.0 ml, SDS 7.5 ml, 2-β-mercaptoethanol 2.5 ml)
- LPS solution of different origin
- Mol weight marker (Rainbow, Amersham)
- bromphenol blue (marker for electrophoresis)

- methanol
- blot-buffer (25 mM Tris, 192 mM Glycin, 10% methanol)
- PBS-Tween 20 [0.1%]
- PBS-Tween 20 (0.1% + 3% skimmed milk, Glücksklee, Nestlé)
- chicken anti-LPS Ab diluted 1:100 in PBS-Tween
- POD-labeled anti-chicken IgY diluted 1:5,000 (SIGMA)
- substrate solution (DAB, see Subprotocol 18 or Chapter 8)

## Procedure

**Step 1 Separation of LPS-samples using a modified continuous system according to Laemmli 1970**

1. Prepare the separation gel (gradient 12-16%, pH 7.4, SDS 1%, stacking gel 4% otherwise as with the separation gel), gel size 140 x 160 x 1 mm.
2. Dilute LPS stock solutions (10 mg/ml) with sample buffer (1:9, vol/vol) and incubate for 5 min at 100°C.
3. Load the LPS sample and the mol weight marker, respectively on the gel in a volume of 20 µl/slot.
4. Run the electrophoresis (40 V, 30 mA/gel, ca. 18 h), under these conditions the bromphenol blue line migrates approximately 10 cm.
5. Keep the temperature during the electrophoresis at 5°C (flow cooler) and exchange continuously the buffer between the upper and lower buffer reservoire (peristaltic pump).

**Step 2 Transfer of the separated substances on the blot membrane modified according to Towbin et al. 1979, and Towbin and Gordon 1984**

1. Treat the blot membrane with methanol for 2 min and wash subsequently with distilled water.

2. Prepare the blot-sandwich according to the instructions of the manufacturer.
3. Run the transfer at RT for 18 h (0.1 mA, 10V).
4. After the transfer rinse the membrane shortly in methanol and dry the membrane for 30 min at RT.
5. For orientation stain the mol weight marker with amido black.
6. Block non-specific binding by incubation in PBS-Tween with 3% skimmed milk for 6 h at RT (shake continuously).
7. Wash thrice with PBS-Tween for 15 min.
8. Incubate in the chicken anti-LPS Ab for 18 h at RT (shake continuously).
9. Wash as above.
10. Incubate in the anti-chicken IgY Ab for 1 h (shake continuously).
11. Wash as above.
12. Stain the membrane with DAB as described in Subprotocol 18 and block the reaction with distilled water.
13. Dry the membrane at RT.

## Results

See Figure 15.

## Troubleshooting

The transfer sandwich must be prepared carefully to avoid air bubbles which can disturb the correct transfer of the substances separated. Otherwise as with Subprotocol 18 (for further technical details see Kyhse-Anderson 1984, Bjerrum and Schaffner-Nielsen 1986).

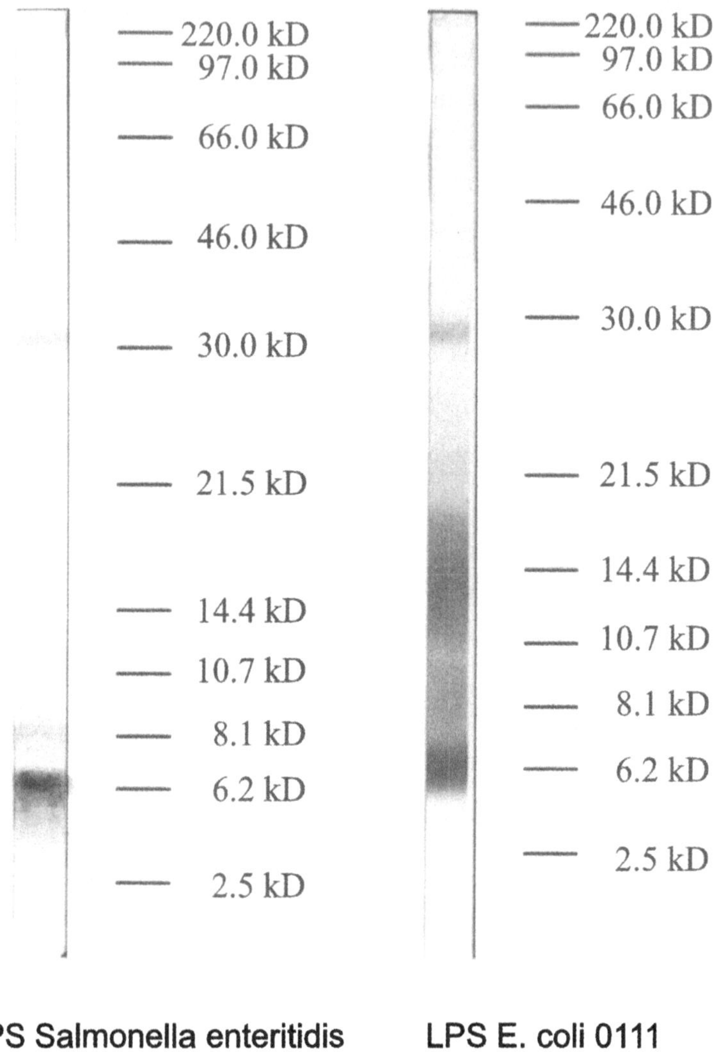


**Fig. 15.** Western blot. Specificity analysis of chicken antibodies raised against lipopolysaccharide (LPS)-species of different origin. *kD* = kilo Dalton.

## Subprotocol 21
## Immunohistochemistry, Visualisation of CCK-8 by an Indirect Staining Method

Immunohistochemical processes are used to reveal tissue-structures. In principle, the assays are designed like an ELISA, so there are both direct and indirect forms. Moreover there are various methods, like the biotin-avidin one, for strengthening the specific signal. For the sake of good results, the tissue to be examined should be suitably prepared, for instance by fixing if necessary. Various procedures are common and may lead to varying results (see Cuello 1983, Ambrosius and Luppa 1987).

### Outline

See Figure 12.

### Materials

**Equipment for step 1**

- beakers
- water-bath
- filtering apparatus

(The solutions should be filtered to leave no foreign bodies visible on the sections.)

**Reagents, solutions for step 1**

- picric acid
- formaldehyde
- mercuric chloride
- glacial acid
- distilled water

**Equipment for step 2**

- 2 peristaltic pumps (e.g. Ismatec MC 360), for a buffer-solution and Bouin's solution respectively
- pressure-bottles
- connecting-tubes
- valves
- scissors and pincettes
- clamp
- button cannula

**Reagents, solutions for step 2**

- 400 mg/ml cloralhydrate (Merck)
- 200 ml haemaccel/PBS-solution (1:5 v/v)
- 500 ml Bouin's solution

**Equipment for step 3**

- vibratome, cryo-microtome
- vials, 5-10 ml (Greiner), covered by fine sieves
- slides
- small brush
- Petri-dishes
- microscopes
- shaker
- magnifying-glass

The fine sieves are for covering the vials in which the sections are freely floating, so that solutions can be poured out without losing the sections.

The brush is used for picking the floating sections up.

**Reagents, solutions for step 3**

- 1% $H_2O_2$-solution in 100% methanol
- PBS
- PBS with 2% goat serum and 0.5% Triton X 100
- PBS with Triton X 100 (0.5%)
- primary IgY-antibody in working dilution

- species-specific POD-labeled secondary antibody in working dilution
- 15 mg DAB (Sigma) in 50 ml PBS with 1500 ml ammonium nickel sulphate (3% w/v) and with 15 ml $H_2O_2$ (1 ml 30% $H_2O_2$ in 30 ml PBS)
- gelatine
- alcohol series
- entallan (Merck)

## Procedure

### Preparation of the fixative

Step 1

There are several possibilities:

- The use of tissue-samples already fixed (often by means of formalin), and the preparation of tissue-sections by a cryo-microtome or vibratome.
- The use of samples already fixed after transfer into paraffin, with the 'paraffin-dependence' of the primary antibody being a problem, and the preparation of tissue-sections by a microtome.
- Intracardial perfusion of an anaesthetized animal (rat) with a solution of fixative, followed by point 1 or 2.

The intracardial perfusion of an anaesthetized animal with Bouin's sublimate solution as a fixative (Sofroniev, 1987) is described below.

For further fixative solutions see Cuello 1983 and Ambrosius and Luppa 1987.

Bouin's sublimate solution is made as follows:

1. Add 1 l boiling distilled water to 200 g picrid acid, to form a saturated solution.
2. Concentrated solution of formaldehyde (37%).
3. Make a saturated aqueous sublimate solution (mercuric chloride, $HgCl_2$ [6% w/v in distilled water], dissolve the sublimate in a boiling water-bath).

4. Glacial acetic acid (to be added just before use).
5. Filter the single solutions.
6. Mix the solutions 1, 2, 3 and 4 just before use in the ratios 15:5:2:1 e.g. 315 ml, 105 ml, 42 ml, 21 ml.

**Step 2 Cardial perfusion of a rat**

1. Anaesthetise a rat with chloralhydrate (400 mg/kg body weight, i.p.).
2. Open the rat's rib-cage.
3. Cut off the tip of the heart and push a button cannula into the left ventricle as far as the aorta and fix the button cannula so that the heart-opening is shut.
4. Clamp off the body circulation with an appropriate clamp (Clamping the Vena cava and the Aorta descendens below the diaphragm cuts the greater circulation off, so the fixative is pumped solely through the pleural/brain circulation).
5. Open the right heart-vestibule, so as to ensure that the fixative can flow out.
6. Perfuse the anaesthetised rat with about 200 ml Haemaccel/PBS-solution (1:5 v/v), to rinse the blood away, using a pressure of 50-100 ml/min applied by an Ismatec MC 360 pump.
7. Perfuse with about 500 ml Bouin's solution, using the same pressure as before.
8. Prepare the brain.
9. Fix once more for 2 h in Bouin's sublimate solution.

**Step 3 Visualisation of the antigen with IgY-primary antibody**

See Figure 12 for an indirect immunohistochemical marking of antigen.

1. Cut the brain using a cryo-microtome or vibratome (we use a vibratome, the thickness of the sections is 50 μm).
2. Put the sections freely floating in PBS.

3. Wash the sections 3 x 15 min freely floating in PBS (in vials, Greiner), then put the vials onto a shaker, as also in the following steps.
4. Incubate the sections for 30 min freely floating in a 1% $H_2O_2$-solution in 100% methanol, repressing the non-specific peroxidase-reaction.
5. Repeat step 1.
6. Incubate for 30 min in PBS with 2% goat serum and 0.5% Triton X 100, blocking the non-specific binding sites.
7. Repeat step 1.
8. Incubate overnight for about 16 h at room temperature in primary IgY-antibody. (In this case we used the antibody Ch 22 diluted 1:500 with PBS-Triton X 100, see Figure 13).
9. Repeat step 1.
10. Incubate for 2 h in the species-specific POD-labeled secondary antibody (from donkey), diluted with PBS-Triton X 100).
11. Repeat step 1.
12. Stain the sections for about 10 min or less in a filtered DAB solution (15 mg DAB in 50 ml PBS), add 1500 μl ammonium nickel sulphate (3% w/v) and add 15 μl $H_2O_2$ (1 ml 30% $H_2O_2$ in 30 ml PBS).
13. Stop the reaction with distilled water.
14. Mount the sections on gelatine slides coated through the following procedure: Dissolve 1 g gelatine (type A or B, Sigma) in 50 ml distilled water boiling in a bath, add a pinch of chromalum, coat the slides with this solution, then dry the slides at 60°C for about 12 h.
15. Dry the sections at room temperature.
16. Transfer the sections into a rising alcohol series and leave them for 10 min in each of the following solutions respectively:
    - 30% alcohol
    - 50% alcohol
    - 70% alcohol

- 90% alcohol
- 96% alcohol, 1
- 96% alcohol, 2
- isopropanol
- xylol

17. Cover the sections with entellan (Merck).

## Results

Fixing the tissues is a decisive step which may influence the results. In determining peptides, we have had good experiences with Bouin's solution, as have other authors (Ambrosius and Luppa 1987, Sofroniew 1983). As second Ab we use an anti-chicken IgY Ab from donkey, Dianova (703-035-155).

See Figure 16 a-f (see also Schade 1991).

## Troubleshooting

Non-specific markings are very problematic. To avoid non-specific staining, it is advisable to dilute the primary or secondary antibodies with PBS, which contains normal serum (about 2%) from other species like goats, sheep and horses.

A precondition for a good result is a good fixation. A critical point is the formaldehyde solution, which may oxidize with increasing usage. In the case of frozen sections, the preservation of tissue-structure can be a problem.

The staining-solution should be prepared just before use and the reaction takes place in the dark. The sections are then blue-black. Take care that the DAB is not to old. Dissolved DAB should have a slight violet gleam.

We have achieved good results with this fixing and staining method. For further variations see Chapter 8, References.

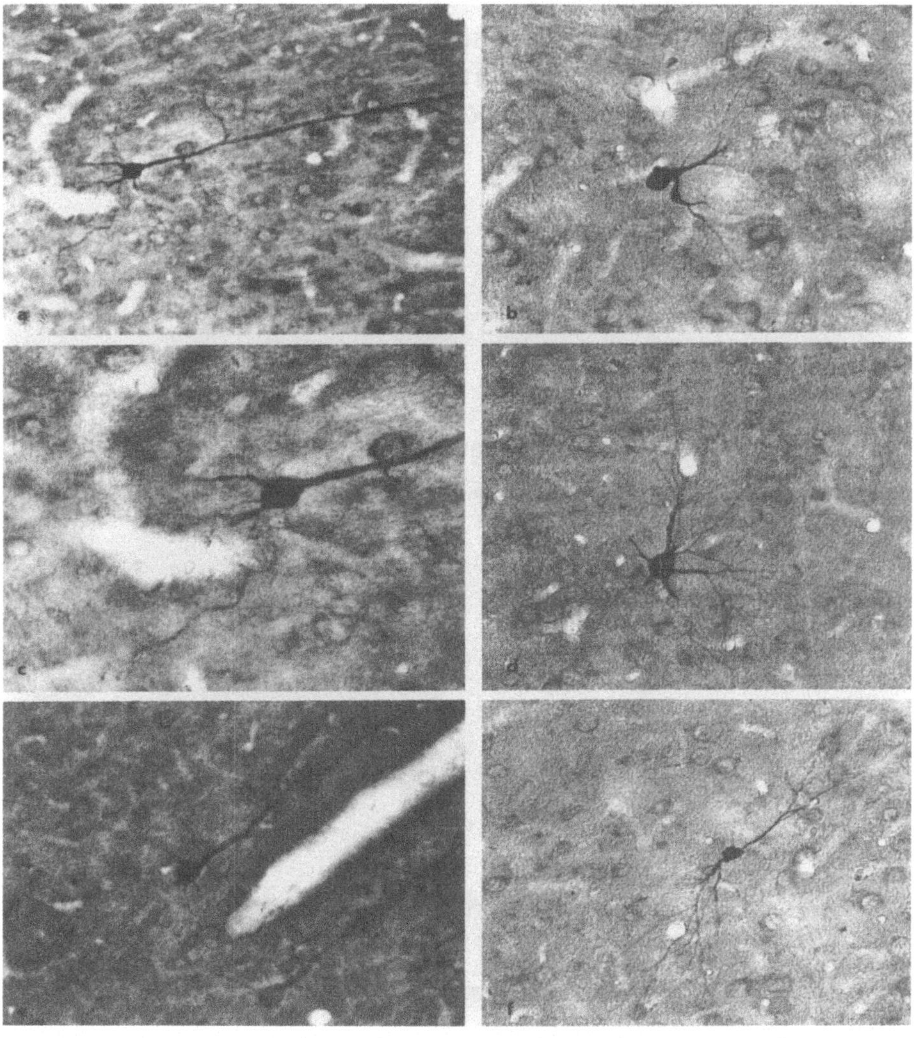

**Fig. 16.** Indirect visualisation of neurons with CCK-like immunoreactivity in the cortex and hypothalamus of the rat brain. As a primary antibody, an IgY anti-CCK antibody (Ch 22) was used. As an antigen, a MAP-CCK8 construct was used, which makes coupling to a protein carrier unnecessary (see Figure 13). As a secondary antibody, a commercial POD-marked antibody from donkey (Dianova) was used. In Figures 15a, c, the pyramidal neurons in the cortex are visualized. In Figure 15e, two neurons are marked with different degrees of intensity. This was not part of the method but is a result due perhaps to various degrees of activity of the neurons. Figures 15b, d, f show three different kinds of neurons from the thalamus of rats. Cortical pyramidal neurons have so far been successfully detected only twice (Sakamoto et al. 1984, Morino et al. 1994). The detection in the former case has been ignored in the relevant literature, and the detection in the latter case succeeded only after preliminary treatment with colchicin and an antibody developed against pro-CCK. We have two IgY antibodies (antigen: CCK8-derivatives) which reliable visualize pyramidal neurons.

## Subprotocol 22
## Immunohistochemistry Detection of MHC Class II-Antigens in Cryostate-Slices From the Lung Lymph-Nodes of Pigs Using Mouse Monoclonal Antibodies

### Outline

See Figure 17.

### Materials

See Subprotocol 21, equipment and materials for step 3, additionally:

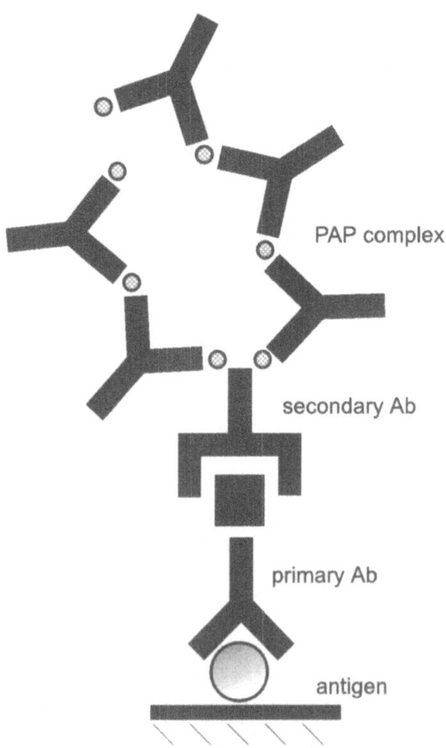


**Fig. 17.** Indirect marking through the PAP-technique (PAP = peroxidase-anti-peroxidase).

- frozen sections of lung lymph-nodes from pig (6-8 μm)
- PBS (pH 7.4)
- mouse mAb (MSA3, 2G6: isotype IgG)
- POD-labeled chicken anti-mouse IgG
- peroxidase-anti-peroxidase-complex (PAP-complex: SIGMA) diluted 1:800
- 3,3'-diaminobenzidine (DAB: Merck): 1 mg/ml PBS
- Canada balsam (Riedel de Häen)

## Procedure

1. Prepare tissue sections as above using a cryostate.
2. Fix the air-dried sections in acetone for 10 min at room temperature.
3. Incubate with primary antibody for 30 min in a moist chamber at room temperature. To safe Ab-solution the sections may be encircled with a DAKO Pen (DAKO). In this way a barrier is formed which keeps the Ab-solution within the encircled area.
4. Wash thrice with PBS.
5. Incubate with POD-labeled secondary antibody (hen anti-mouse immunoglobulin) for 30 min in a moist chamber at room temperature.
6. Wash thrice with PBS.
7. Incubate with the PAP-complex.
8. Wash thrice with PBS.
9. Visualise the enzyme-linked antibody by having it react with DAB and hydrogen peroxide as described above and in Subprotocol 21 at room temperature for 10 min.
10. Counterstain the sections slightly with haematoxylin.
11. Mount them with Canada balsam.
12. Examine them by light microscopy.

## Results

See Figure 18.

## Troubleshooting

See Subprotocol 21.

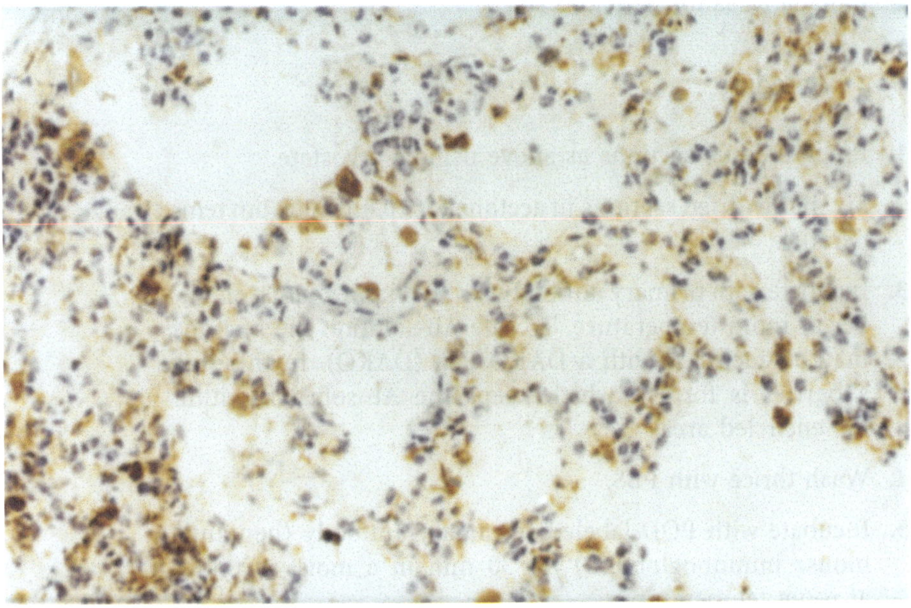

**Fig. 18.** Demonstration of MHC class II antigens on macrophages (lung lymph-nodes of a pig) according to Subprotocol 22. The brown colored cells are MHC class II positive.

## Subprotocol 23
## Detecting GABAergic (g-Amino-Butyric Acid) Cells in the Brain of Mongolian Desert Mice (*Meriones ungulatus*)

### Outline

In this assay the chicken Ab is used as a non-labeled species-specific secondary Ab which is detected by a POD-labeled tertiary rabbit anti-chicken Ab.

### Materials

See Subprotocol 21, additionally:

- brain sections (cryostate, 50 µm)
- methanol, PBS, goat normal serum, BSA, Triton X-100, sodium azide, Tris buffer
- mouse anti-glutamic acid decarboxylase (GAD, Boehringer Mannheim)
- unlabeled chicken anti-mouse Ig, diluted 1:15
- rabbit anti-hen IgY (POD-labeled, Chemicon, Temecula) diluted 1:500
- DAB solution (0,02%) in TBS with 0.05% $H_2O_2$

### Procedure

1. Take the sections free floating in PBS in multi-well plates.
2. Incubate the sections for 60 min at 4°C in a 1% $H_2O_2$ solution in PBS/methanol (1:1), suppressing the non-specific peroxidase reaction.
3. Wash the sections 3 x 20 min in PBS.
4. Incubate the sections for 60 min at room temperature in PBS with 2% goat normal serum and 5% BSA, 0.1% Triton X 100, 0.01% $NaN_3$ blocking the nonspecific binding sites.

5. Incubate for 2 days at 4°C with anti-GAD (diluted 1:5.000 in PBS, 5% BSA, 0.1% Triton X 100, 0.01% $NaN_3$).
6. Wash the sections 3 x 20 min in PBS.
7. Incubate for 150 min with chicken anti-mouse IgG (diluted 1:15 in PBS, 1% BSA, 0.01% $NaN_3$).
8. Wash the sections 2 x 20 min in PBS and 1 x 20 min in TBS.
9. Incubate for 60 min at room temperature with POD-labeled rabbit anti-chicken IgY (diluted 1:500 in TBS, 1% BSA).
10. Wash the sections 3 x 20 min in TBS.
11. Stain the sections for 15 min in DAB solution.
12. Wash the sections 2 x 30 min in TBS and overnight in PBS at 4°C.
13. Mount the sections on slides.
14. Dry the sections at room temperature.
15. Transfer the sections into a rising alcohol series and leave them for 2 min in each of the following solutions respectively: optal, xylol (2 x) (see also Subprotocol 21).
16. Cover the sections with entellan (Merck).

## Troubleshooting

See Subprotocol 21.

# Subprotocol 24
# Immunocytochemical Detection of CD44 in Human Keratocytes in vitro

## Outline

For principle see Figure 12.

## Materials

See Subprotocol 21, additionally:

- primary culture praeputium, about 7500 cells/well in 16-chamber slides (The culture is kept at 37°C till subconfluence)
- hen normal serum
- TBS, BSA
- anti-CD44 (mouse monoclonal antibody J173, Immuno-Tech)
- Cy3-labeled hen anti-mouse Ig
- entallan (Merck)
- fluorescence microscope

## Procedure

1. Keep the cells at 37°C till confluence.
2. Fix the cells for 9 min with methanol, for 1 min with aceton (-20°C) on slides.
3. Dry the slides.
4. Wash the slides for 10 min in 0.05 M TBS pH 7.6.
5. Block with hen normal serum (0.1% in TBS with 5% BSA) for 2 h at room temperature.

6. Add the primary Ab, anti-CD 44, diluted 1:30 in TBS with 5% BSA.
7. Incubate the slides overnight in moist atmosphere at 4°C.
8. Wash the slides 3 x 10 min in 0.05 M TBS.
9. Add the secondary antibody Cy3-labeled hen anti-mouse immunoglobulin diluted 1:100 in TBS with 5% BSA.
10. Incubate the slides overnight in moist atmosphere at 4°C.
11. Wash the slides 3 x 10 min in 0.05 M TBS.
12. Transfer the slides in a rising alcohol series (see Subprotocol 21, step 3).
13. Cover the slides with entallan (Merck).
14. Analyse the slides by means of a fluorescence microscope.

## Troubleshooting

See Subprotocol 21.

## Subprotocol 25
## Use of IgY-Ab as Secondary Antibodies in Flow Cytometry: Immunofluorescent Staining of a Single Cell Suspension from Human Peripheral Blood to Detect Surface Antigens Through One-Colour Labeling (Indirect Method)

The following reactions show the use of hen antibodies as secondary ones, as well as the efficacy of IgY-antibodies as double ones (labeled and unlabeled) and as double sandwich ones. Flow cytometry is widely used for detecting the expression of cell-surface and intracellular molecules and is used mainly to measure the intensity of fluorescence.

Antibodies fluorescently-labeled can bind specifically to soluble antigens or cell-associated molecules. The direct and indirect immunofluorescence of cells can yield information about the antigen and the binding of antibodies. The antigen-antibody complexes can be detected by means of secondary antibodies coupled to fluorescent stains.

Flow cytometry data are generally analysed to type and define different cell populations and/or to ascertain the relative amounts of cell-surface antigen-molecules.

The frequency is determined by the analysis of single parameter fluorescence (histograms, see Fig. 20) and of dual-parameter fluorescence (dot points, see Fig. 22). The direct immunofluorescent method is faster but less sensitive than the indirect one. Many distinct antibodies (e.g. mouse ones) can be detected with the latter.

In indirect staining the primary reagent is not labeled but detected by a second fluorochrome-labeled reagent. This second reagent may be an antibody reacting specifically to the first one. Alternatively the streptavidin/avidin system can be used, where an antibody is conjugated to biotin and detected with phycoerythrin-labeled streptavidin/avidin.

## Outline

See Figure 19.

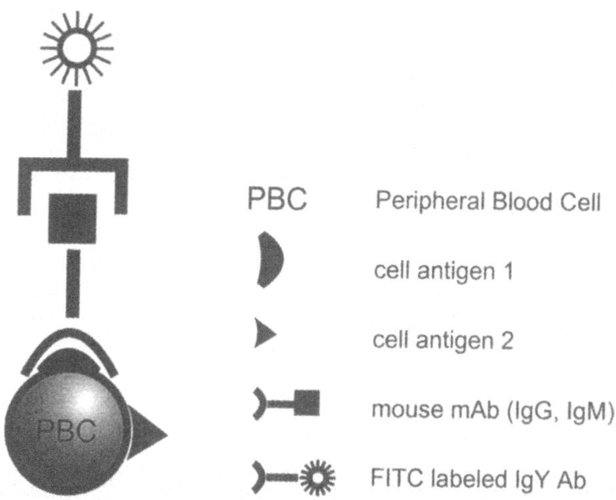


**Fig. 19.** Scheme of indirect marking of peripheral blood cells (PBC) for FACS analysis.

## Materials

**Equipment**

- polystyrene microtitre plates (96, V)
- pipette and pipette tips
- centrifuge with rotor for microtitre plates
- vortexer
- FACscan (Becton Dickinson)
- refrigerator

**Reagents, solutions**

- living cells, like peripheral human lymphocytes
- PBS
- PBS/0.01 M sodium azide (washing solution)
- antibodies:
  - mouse monoclonal antibody in appropriate dilution
  - FITC-labeled hen anti-mouse IgY in appropriate dilution

## Procedure

1. Add 50 µl unlabeled mouse antibody to the well on the plate.
2. Add 50 µl cell suspension (2-10 x $10^6$/ml)
3. Mix gently.
4. Incubate for 45 min at 4°C in a moist atmosphere.
5. Wash the cells thrice with a wash buffer (50 µl, 100 µl, 100 µl) at 2 min intervals and each time centrifuge at 300 g for 3-5 min after washing.
6. Stain the cells with 50 µl FITC-labeled hen anti-mouse Ig.
7. Incubate for 30 min at 4°C in a moist atmosphere.
8. Wash the cells thrice as above.
9. Fix the cells, using 100 µl 1% formaldehyde/wash buffer.
10. Keep the cell suspension at 4°C until analysed by flow cytometry.

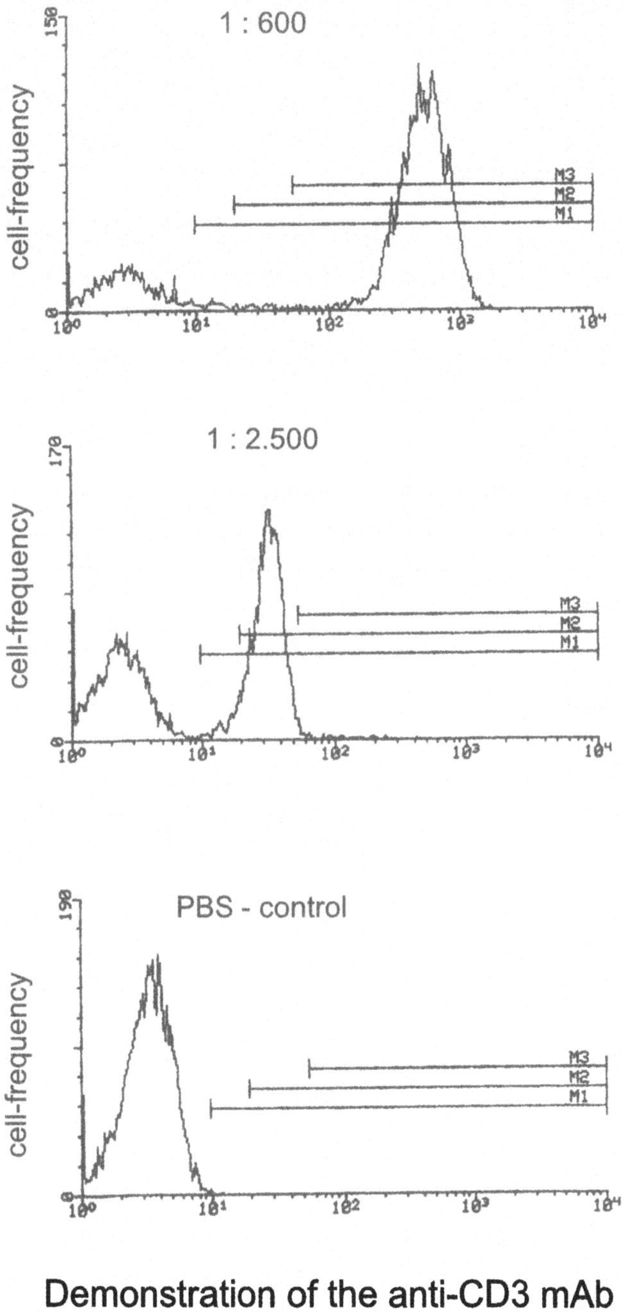


**Fig. 20.** Histogram of FACS-analysis to detect mouse mAB using biotin- or FITC-labeled chicken anti-mouse antibodies (see also Figure 21).

## Results

See Figure 20.

## Troubleshooting

Note that there is no loss of cells during the wash procedures. The resuspension after washing is performed merely by gentle lateral knocking on the microtiter plates. The analysis of the labeled and fixed cells can be performed in one week.

# Subprotocol 26
# Double-Immunofluorescent Staining of a Single Cell Suspension From Human Peripheral Blood to Detect Surface Antigens Through Two-Colour Labeling (Indirect/Direct Method)

## Outline

See Figure 21.

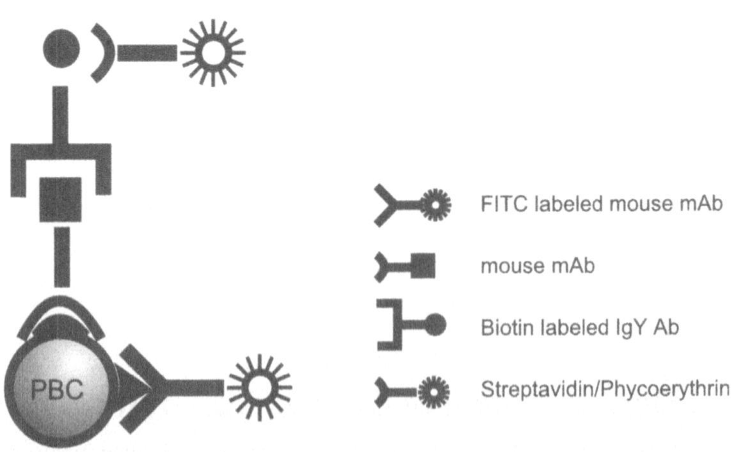


**Fig. 21.** Scheme of direct/indirect marking of PBC for FACS analysis.

## Materials

See Subprotocol 25 (except antibodies).

**Antibodies**

- unlabeled mouse antibody (IgM, directed against CD4, B40, Ia/1) in appropriate dilution
- biotin-labeled hen anti-mouse IgM/streptavidin-phycoerythrin
- FITC-labeled mouse antibody (IgG, directed against CD3) in appropriate dilution

## Procedure

1. Add 50 µl unlabeled mouse antibody (IgM) to the wells on the plate.
2. Add 50 µl cell suspension (2-10 x $10^6$/ml).
3. Mix gently.
4. Incubate for 45 min at 4°C in a moist atmosphere.
5. Wash the cells thrice with a wash buffer (50 µl, 100 µl, 100 µl) at 2 min intervals and each time centrifuge at 300 g for 3-5 min after washing.
6. Suspend the cell-pellet again gently.
7. Incubate the cells with 50 µl hen-anti mouse IgM-biotin.
8. Incubate for 30 min at 4°C in a moist atmosphere.
9. Wash as above.
10. Add 50 µl labeled streptavidin-phycoerythrin.
11. Incubate for 30 min at 4°C in a moist atmosphere.
12. Wash as above.
13. Add 50 µl FITC-labeled monoclonal mouse antibody (isotype IgG).
14. Incubate for 30 min at 4°C in a moist atmosphere.
15. Wash as above.

16. Fix the cells, using 100 μl 1% formaldehyde/wash buffer.
17. Keep the cell-suspension at 4°C until analysed by flow cytometry.

## Results

See Figure 22.

## Troubleshooting

See Subprotocol 25.

# Subprotocol 27
# Double-Immunofluorescent Staining of a Single Cell Suspension from Human Peripheral Blood to Detect Surface Antigens Through Two-Colour Labeling (Indirect/Indirect Method)

## Outline

See Figure 23.

## Materials

**Equipment** See Subprotocol 25.

**Reagents, solutions** See Subprotocol 25 (except antibodies, specificity of the mouse mAb as in Subprotocols 25 and 26).

**Antibodies**

- unlabeled mouse monoclonal antibody (different isotypes, e.g. IgG1, IgG2b or IgM)
- FITC-labeled hen-anti mouse IgG1
- biotin-labeled hen-anti mouse IgG2b or IgM
- streptavidin-phycoerythrin in appropriate dilution

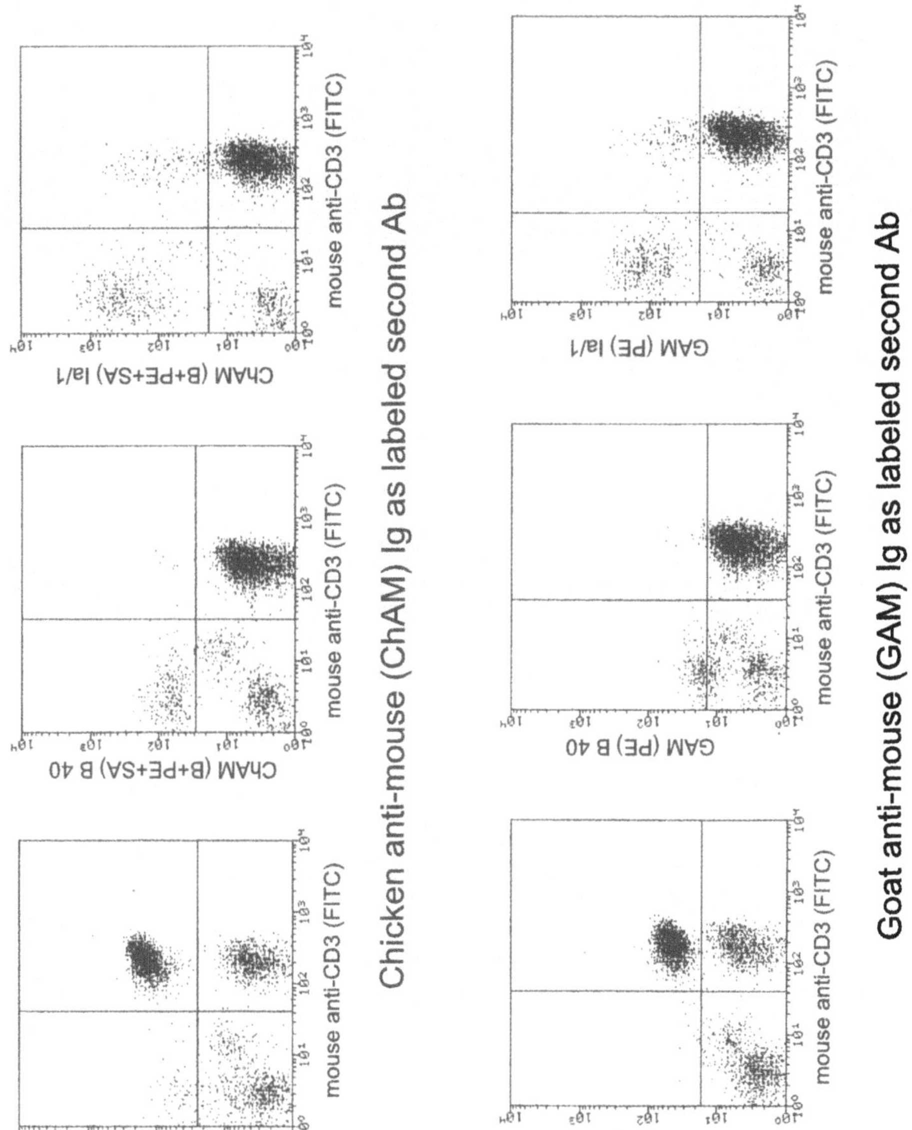


**Fig. 22.** Dot-plot of FACS-analysis to demonstrate several human cell differentiation antigens by means of the direct/indirect method. Comparison between the efficacy of an avian and a mammalian antibody.

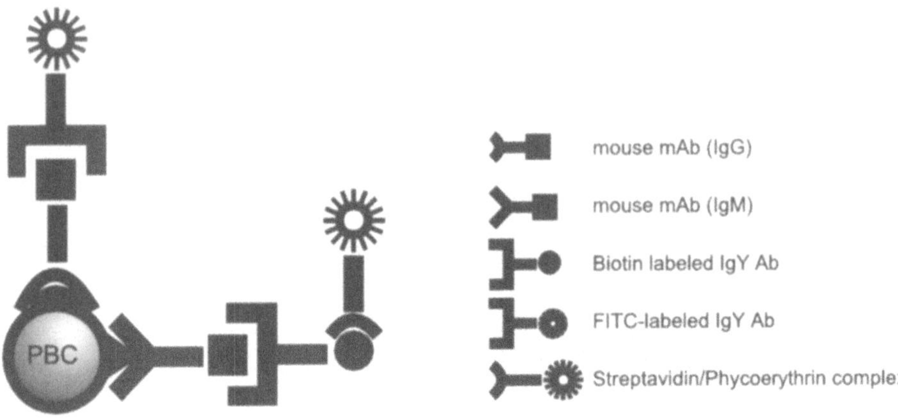


**Fig. 23.** Indirect/indirect marking of PBC for FACS-analysis.

## Procedure

1. Add 50 µl unlabeled mouse mAb (IgG1) to the wells on the plate.
2. Add 50 µl cell suspension (2-10 x $10^6$/ml).
3. Mix gently.
4. Incubate for 45 min at 4°C in a moist atmosphere.
5. Wash the cells thrice with a wash buffer (50 µl, 100 µl, 100 µl) at 2 min intervals and each time centrifuge at 300 g for 3 min after washing.
6. Suspend the cell pellet again gently.
7. Stain the cells with 50 µl FITC-labeled hen-anti mouse IgG1.
8. Incubate for 30 min at 4°C in a moist atmosphere.
9. Wash as above.
10. Add 50 µl unlabeled mouse mAb (IgG2b or IgM).
11. Incubate for 30 min at 4°C in a moist atmosphere.
12. Wash as above.
13. Add 50 µl biotin-labeled hen-anti mouse IgG2b or IgM.
14. Incubate for 30 min at 4°C in a moist atmosphere.

15. Wash as above.
16. Add 50 μl streptavidin-phycoerythrin.
17. Incubate for 30 min at 4°C in a moist atmosphere.
18. Wash as above.
19. Fix the cells, using 100 μl 1% formaldehyde/wash buffer.
20. Keep the cell-suspension at 4°C until analysed by flow cytometry.

## Troubleshooting

See Subprotocol 25.

## Subprotocol 28
## Competitive Radioimmunassay Determination of Neuronal CCK-8 (Cholecystokinin Octapeptide)

In radioimmunoassays a reaction-partner is radioactively marked, and here the antigen (CCK) is marked with $^{125}I$. By measuring the radioactivity of the reaction-product (Ag-Ab complex), conclusions can be drawn about the amount or concentration of the substance investigated. As with ELISA there are competitive and non-competitive assays. In the former, a marked and a non-marked substance compete for a limited number of binding sites, and in the latter a substance reacts with a marked binding-protein.

## Outline

See Figure 24.

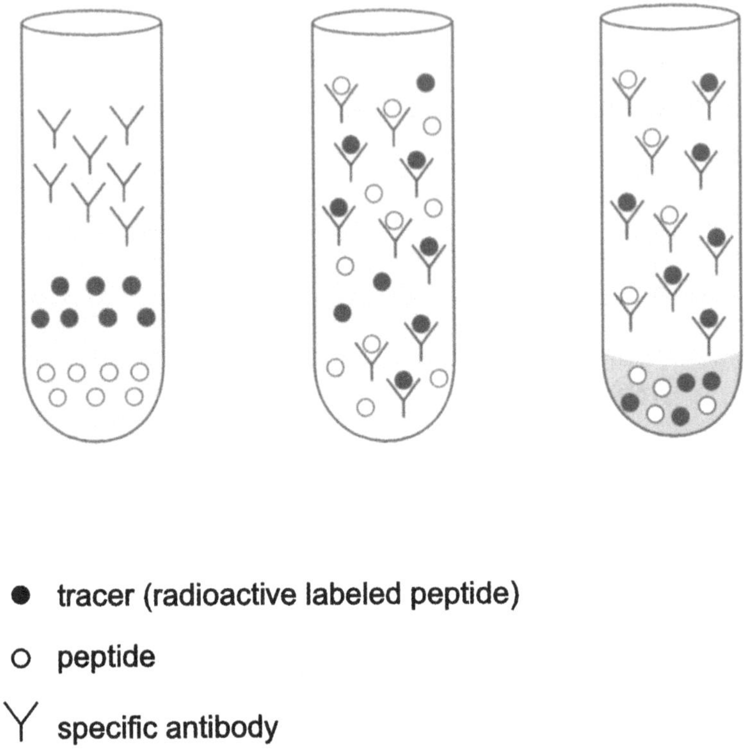


**Fig. 24.** A competitive radio immunoassay, taking as an example a peptide (CCK)-determination. The amount of the antibody (IgY anti-CCK antibody) and of the tracer are defined. The concentration of the samples varies. The more CCK contained by a sample, the less the marked tracer can be bound, and the lower is the sample's radioactivity.

## Materials

**Equipment**

- vortex
- refrigerator
- cool-centrifuge (e.g. Hermle Z 360 K, rotor 220.27 V03)
- g-counter (e.g. 1470 Wizard, Pharmacia)
- pipettes
- pipette-tips
- safety-equipment for radioactive-work

**Reagents, solutions**

- RIA buffer: see Chapter 8
- plastic tubes (volume ca. 1 ml, compatible with the g-counter used)
- samples and standard (see Table 2)
- tracer (radio-labeled, adjusted to 1.000 cpm/100 ml)
- anti-CCK IgY-antibody diluted with RIA buffer (1:1.000)
- charcoal (Norit A, Serva) in RIA buffer

## Procedure

1. Prepare the IgY-antibody dilution with RIA buffer.
2. Add the following solutions in a plastic tube (see Table 3):
   - Standard or sample
   - RIA buffer
   - Tracer ($^{125}$I-CCK)
   - Diluted IgY-antibody
3. Shake the mixture, using a vortex.
4. Incubate for 24 h or longer (see Chapter 8) in a refrigerator at 4°C.
5. Separate the bound from the unbound tracer by adding 500 µl of the charcoal mixture to the solutions in the tube.
6. Shake vigorously.
7. Incubate for 1 h in a refrigerator at 4°C.
8. Centrifuge the tubes for 20 min in a cooling- centrifuge at 4,000 cycles/min.
9. Decant the supernatant completely, pouring it out into another plastic tube.
10. Measure the supernatant with a γ-counter for 3-10 min.

**Table 2.** Example of a standard curve

| | | |
|---|---|---|
| Standard 6 | 0.2 fMol/100 µl | 100 µl |
| Standard 5 | 1.0 fMol/100 µl | 50 µl |
| Standard 5 | 2.0 fMol/100 µl | 100 µl |
| Standard 5 | 4.0 fMol/100 µl | 200 µl |
| Standard 4 | 10.0 fMol/100 µl | 50 µl |
| Standard 4 | 20.0 fMol/100 µl | 100 µl |
| Standard 3 | 80.0 fMol/100 µl | 100 µl |

The various volumes are equalised by means of the RIA buffer (Standard + RIA buffer = 200 µl). The standard curve is calculated with the RIA Calc Program (LKB, Wallac) as is the precision profile, which yields information about the most precise area of measurement (see Figure 25).

**Table 3.** Example of an RIA pipetting-scheme on the basis of an IgY-antibody

| | Standard or sample | RIA-buffer | Tracer $^{125}$I-CCK | Antibody | Charcoal |
|---|---|---|---|---|---|
| Non-specific binding | - | 300 µl | 100 µl | - | 500 µl |
| Specific binding | - | 200 µl | 100 µl | 100 µl | 500 µl |
| Total | - | 800 µl | 100 µl | - | - |
| s6 | 100 µl | 100 µl | 100 µl | 100 µl | 500 µl |
| s5 | 50 µl | 150 µl | 100 µl | 100 µl | 500 µl |
| s5 | 100 µl | 100 µl | 100 µl | 100 µl | 500 µl |
| s5 | 200 µl | | 100 µl | 100 µl | 500 µl |
| s4 | 50 µl | 150 µl | 100 µl | 100 µl | 500 µl |
| s4 | 100 µl | 100 µl | 100 µl | 100 µl | 500 µl |
| s3 | 100 µl | 100 µl | 100 µl | 100 µl | 500 µl |

## Results

Through addition of active charcoal, the bound radioactivity is separated from the unbound (see Figure 24). Other means of separation are possible (e.g. separation by means of species-spe-

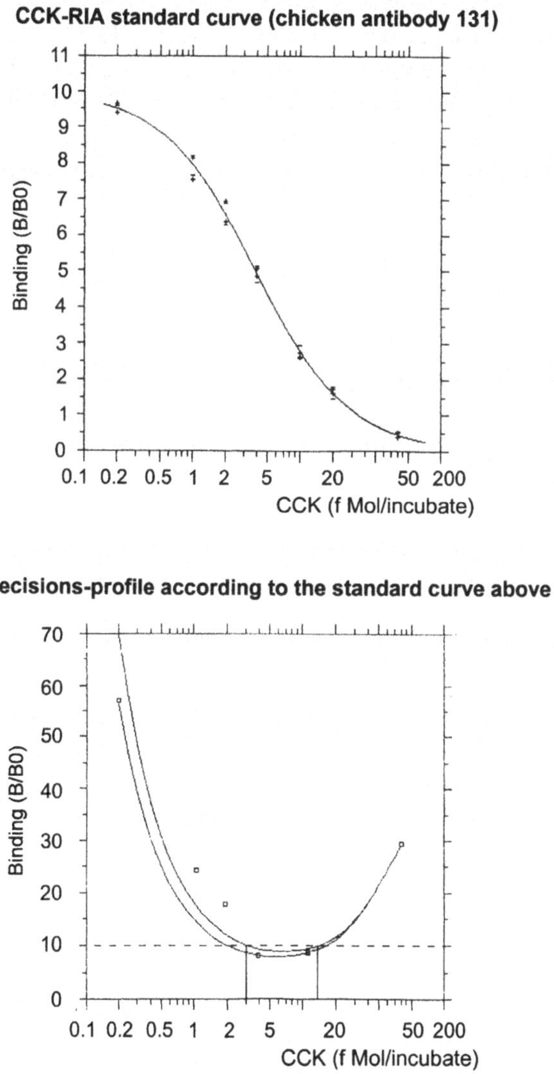


**Fig. 25.** An example of a standard curve (upper part). In the lower part is the corresponding precision-profile, which shows in which range measurements are most reliable. Here it is from about 3-15 fMol/incubate.

cific antibodies). For standard curve and precision profile see Figure 25 (Schade et al. 1988, Schade et al. 1996, Vick 1995).

## Troubleshooting

There may be an increasing non-specific binding with increasing age of the tracer (decay period). Depending on the quality of the antibody used, an incubation-period of more than 24 h may be needed. A pre-incubation of antibody and sample may raise the sensitivity of the sample. A good separation of the bound from the unbound tracer may depend on the charcoal used.

# Subprotocol 29
# Complement Fixation Test (CFT)

In its latest edition of the O.I.E. "Manual of Standards for Diagnostic Tests and Vaccines" ("Office International des Epizooties", Paris 1996), CFT is second only to ELISA as a prescribed or recommended way of diagnosing infections. Its reliability, in diagnosing brucellosis for instance, has established it as a confirmatory test.

Complement (C) belongs to the non-specifically acting defense mechanisms which by itself doesn't act as an antibody but is needed for destruction of antigens by increasing phagocytosis or by extracellular lysis. In serum, the complement system (C1 - C 9 with additional subfactors) is present in an inactive form. Activation of C is caused by the formatiòn of immune complexes (IC) where the first C-fraction "C1q" joins the IC and thereby activates the cascade of subsequent C-fractions.

## Outline

In vitro, the CFT provides an answer to the question whether C has been fixed by an antibody-antigen complex (IC) (positive case) or not (negative case). It is obvious that the quantity of C plays an important role in regard to the outcome of the test. Therefore, C from the side of the test-serum has to be inactivated by heating to 56 - 60°C. For the test, heat inactivated C is replaced by a foreign properly diluted C collected normally from guinea-pig.

The test runs in two steps: During the first step, C is either fixed to the IC (positive), or not (negative), and is consequently either not available anymore or still available. During the second step, the resulting availability is tested by a second IC where sheep red blood cells (SRBC) together with anti-SRBC ("amboceptor" or "anti-sheep-hemolysin") form the hemolytic or indicator system. In a positive case, no C is available to cause lysis - here hemolysis- during the second step, and vice versa (Figure 26).

After binding to antigens, antibodies from hens are unable to activate the first factor (C 1) of guinea-pig complement and release the C-cascade (Benedict and Yamaga in Marchalonis J: Comparative Immunology, Blackwell Scient Publ 1976). During research work on CFT with avian sera, it was found that the simplest way to overcome this handicap was to add normal hen-serum to guinea-pig C which caused lysis of sensitized SRBC. Apparently, the corresponding C-components of hens and guinea-pigs are compatible (Brumfeld et al. 1961, Marquardt and Newman 1971).

Anti-complement reactions, in part very disturbing, led to further research. Stolfi et al. (1971) used an amboceptor from hens, guinea-pig C minus C 1, and C 1 from chicken to supplement the C1-devoid guinea-pig C. This modified CFT has been applied by Van Regenmortel and Burckhard (1985) for detection of IgY-antibodies directed against rotavirus and plant-virusses ("tomato bushy stunt virus" and "tobacco mosaic virus").The procedure was optimized by adjustment of the dilution-buffer during the procedure.

## Materials

**Equipment**

- photometer (for photometric measurement of hemolysis or for subjective determination of the amount of the non-lysed erythrocytes)
- refrigerator

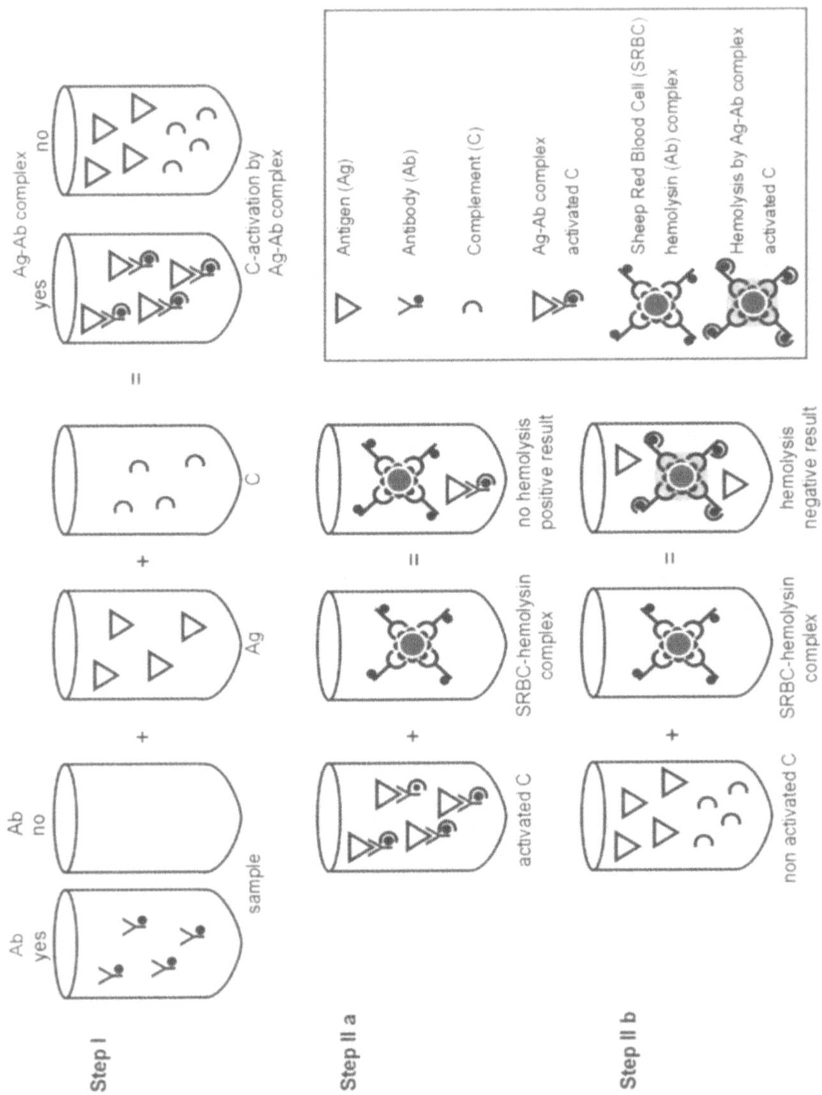


**Fig. 26.** The scheme of the complement fixation test. Stage I represents the stage of complement activation through the formation of the Ag-Ab complex (IC). If there are no antibodies in the sample to be determined, there is also no complement binding. Stage IIa shows full complement binding, so that no C can bind to the SRB hemolysin complex and consequently no hemolysis appears. This means that the sample contains the sought antibody. In stage IIb, there is unbound C, which binds to the SRBC-hemolysin complex and causes hemolysis. This means that the sample does not contain the sought antibody.

- micro-titre plates
- pipettes
- centrifuge
- waterbath 37°C
- waterbath 56 - 60°C
- CFT-buffer

**Reagents, solutions**

- antigen
- antibody
- sheep red blood cells (SRBC)
- amboceptor (antibodies against SRBC)
- complement (from guinea-pigs and hens)

Yolk-antibodies were extracted in the way described by Polsen et al. (1980).

**Buffers**

- veronal-buffer (VB) pH 7.3 - 7.4

| | |
|---|---|
| 85.000 g | NaCl |
| 3.750 g | sodium 5.5-diethyl barbiturate |
| 0.220 g | $CaCl_2 \cdot 2\ H_2O$ |
| 1.015 g | $MgCl_2 \cdot 6\ H_2O$ |
| 5.750 g | 5.5-diethyl barbituric acid |

Dissolve the acid hot and add the other components, then top the mixture up to 2000 ml (dist. water) to produce a fivefold concentrated buffer.

### Preparation of the components

**A1: GVB (gelatin veronal buffer)**

Dilute concentrated VB 1:5 and add 0.1% gelatin.

**A2: DGVB (dextrose GVB)**

Mix equal volumes of GVB and a 5% solution of dextrose in water, while doubling the concentrations of $Ca^{++}$ and $Mg^{++}$.

**A3:DG (dextrose-gelatin)** Mix 5% dextrose in dist. water with 0.1% gelatin, 0.00015 M $Ca^{++}$ and 0.0005 M $Mg^{++}$.

**B: Hemolytic system**

**B1: IgY anti-SRBC Ab (Amboceptor)** Immunise a hen several times with washed SRBC (see Chapter 3), collect serum, mix it with an equal amount of glycerol and store it at -20°C.

**B2: Sensitized SRBC ($Eab^{ch}$)** Take washed sheep blood-cells (SRBC $1x10^8$) in 0.01 M isotonic EDTA in GVB without $Ca^{++}$ and $Mg^{++}$, and sensitize with the same volume of non-activated IgY anti-SRBC diluted 1:100 in the same buffer and incubated for 30 min at room temperature. Wash the sensitized cells and adjust them to a concentration of $10^8$/ml with dextrose GVB for immediate use.

**C1: Chicken C 1** Precipitate the C 1-component of hen-C from the whole serum by addition of 1 volume of serum to 8 volumes 0.02 M acetate buffer, pH 5.0, then dissolve the precipitate in tenfold the original volume of GVB.

**C2: Guinea-pig C without C 1** Resuspend lyophilized guinea-pig C in GVB pH 7.4 then dialyse it against GVB 7.4 ( I=0.04). Remove the resulting precipitate by centrifuging and discard it. The supernatant fluid contains all the C-components of guinea-pig serum save C 1. Test for the optimal concentrations of guinea-pig C minus C 1, of hen C 1 and of hemolysing IgY anti-SRBC with a checkerboard system.

## Procedure

The test is performed in tubes kept in an ice bath. The following reagents, all diluted with GVB were added:

1. 0.2 ml of antigen, 0.2 ml of chicken C1 and 0.2 ml of specific IgY (specific against the antigen) at different dilutions, incubation overnight, 4°C.
2. The ionic strength of the mixture was lowered to I=0.075 by addition of 0.6 ml of DG (to preserve maximal C 1 activity).

3. 0.2 ml of freshly prepared sensitized SRBC were added, incubation at 30°C for 30 min.
4. Addition of 0.2 ml guinea-pig C devoid of C1, incubation for 2 hours, centrifugation and measuring the supernatant by photometer at wavelength 413 nm.

**Controls**

The following controls may be used too, see subsequent diagram, Table 4:

- antiserums in suitable dilutions + C without antigen
- antigen + C without antiserum
- two tubes for single and double C-doses without antigen or antiserum, so as to reveal the C-dose for complete hemolysis
- two tubes with dilution buffers, to measure or exclude spontaneous hemolysis

## Results

Centrifuge the tubes and determine the absorbance photometrically at 413 nm, then express the result as% fixed C through the following formula:

[($A_{413}$ total lysis - $A_{423}$ background) - ($A_{413}$ test - $A_{413}$ background)] : [($A_{413}$ total lysis) - ($A_{413}$ background) x 100]

The CFT is used only for research purposes but not for routine diagnostic work.

## Troubleshooting

Using this test system it is very important to carry out careful controls according to Table 4 and to test out appropriate dilution ratios of the several components involved in reaction.

**Table 4.** Control schedule (Data in ml)

| | IgY-Ab dilution (several) | Chicken C1 | Antigen | Single dose C1 | Double dose C1 | Buffer |
|---|---|---|---|---|---|---|
| 1. | 0.2 | 0.2 | - | - | - | 0.2 |
| 2. | - | 0.2 | 0.2 | - | - | 0.2 |
| 3. | - | - | - | 0.2 | - | 0.4 |
| 4. | - | - | - | - | 0.2 | 0.4 |

## Subprotocol 30
## Field Studies of the Prophylactic use of Specific Egg-Antibodies Against Diarrhoea due to Various Infections in New-Born Calves

### Yolk antibodies in the prophylaxis and therapy of diarrhoea in domestic animals

Diarrhoea due to infection is one of the commonest dangers to animal-life world-wide. The standard therapy is with antibiotics, but the side-effects are notably negative, and the main ones are often negative too. Antibiotic therapy may harm the balance of bacterial species in the intestines, leaving the latter with a predisposition towards further infections, so it should be used to treat only feverish general ailments. In treating diarrhoea, priority should be given to rehydration and electrolytic supply.

Moreover, as a veterinary prophylaxis against diarrhoea in suckling calves and piglets, the dams are now given vaccinations, which via colostrum induce local passive immune protection in the sucklings' guts. Immune protection there is also significantly increased by means of specific antibodies administered orally. The sources of such antibodies may be serum, colostrum, milk and monoclonal antibodies.

An alternative source may be hens' eggs (Lösch et al. 1986), since serum IgY antibodies produced after immunization are transferred through egg-follicles to yolks (Rose et al. 1974, Kowalczyk et al. 1985). Hence IgY from yolk can be produced easily and abundantly, with a yield of up to 250 mg per hen's egg (Erhard et al. 1997). Akita and Nakai (1993) described a total IgY of 8 - 20 mg IgY/ml yolk with up to 10% of specific IgY. Specific IgY is also able to bind or to neutralize pathogens.

The efficacious use of colostrum- and milk-antibodies against diarrhoea had already been described, so it seemed advisable to investigate the efficacy of specific yolk-antibodies in prophylaxis and therapy. Due to the phylogenetic gap between birds and mammals the spectrum of antibodies may be quite different. Birds' may deal with epitopes which have no effects as antigens on mammals.

To check the practical efficacy of orally administered egg-antibodies, the investigations described under the following Subprotocols were carried out.

## Outline

**Experiment 1**

- Investigation of 2 experimental groups:
  - Group 1: E coli K99- and rotavirus-specific egg-powder (antibody-powder 22 g)
  - Group 2: control egg-powder from non-immunized hens (control-powder 22 g)
- Treatment with 22 g of egg-powder per day in the first 14 days post natum (p.n.) twice a day with the meal.

**Experiment 2**

- Investigation of 3 experimental groups:
  - Group 1: E coli K99- and rotavirus-specific egg-powder (antibody-powder 16 g)
  - Group 2: E. coli K99- and rotavirus-specific egg-powder (antibody-powder 8 g)
  - Group 3: without feed additive (control)
- Treatment with altogether 16 or 8 g of egg-powder per day in the first 10 days p.n. twice a day with the meal.

**Experiment 3**

- Investigation of 4 experimental groups:
  - Group 1: E coli K99- and rotavirus-specific egg-powder (antibody-powder 8 g)
  - Group 2: E coli K99- and rotavirus-specific egg-powder (antibody-powder 4 g)
  - Group 3: E coli K99- and rotavirus-specific egg-powder (antibody-powder 2 g)
  - Group 4: without feed additive (control)

- Treatment with altogether 8, 4 or 2 g of egg-powder per day in the first 14 days p.n. twice a day with the meal.

## Procedure

In prophylaxis with specific egg-antibodies, attention has to be paid to not only the range but also the times of possible infection. Calves most often have diarrhoea in the first two weeks of life. The main pathogens are rotaviruses, coronaviruses, K99-positive enterotoxic E. coli and cryptosporides. The pathogen diagnosis was normally carried out on day 7, or in the case of diarrhoea, with the help of Lactovac[8]-ELISA (Hoechst Unterschleißheim, Germany). The parameters considered are the rate, intensity and duration of diarrhoea, as also the gain in body-weight. Altogether, three investigations were carried out independently of one another. In the egg-powder, titres of antibodies were measured against E. coli K99 of at least 330 ELISA-units and against rotaviruses of at least 800 units by the neutralization test. The egg powder was fed with the colostrum or milk.

1. Immunization of laying hens (see Chapter 3).
2. Obtaining egg-powder by spray-drying.
3. Testing the powder for pathogen-specific yolk-antibodies for use in calves and piglets.
4. Pathogen diagnosis of the infected animals.
5. Prophylactic treatment of the infected animals by applying the egg-powder (together with colostrum or milk).
6. Monitoring the development of the infection by means of various parameters (rate, intensity and duration of diarrhoea, body weight gain).

## Results

The field trials of the prophylactic use of hen-antibodies on calves were carried out in three different places: south-west Germany, south-east Germany and Turkey.

**Table 5.** Prophylactic use of specific egg-antibodies against diarrhoea due to various infections in new-born calves

| Parameter | 22g Ab (trial 1) | 22g Co (trial1) | 16g Ab (trial 2) | 8g Ab (trial 2) | Contr. (trial 2) | 8g Ab (trial 3) | 4g Ab (trial 3) | 2g Ab (trial 3) | Contr. (trial 3) |
|---|---|---|---|---|---|---|---|---|---|
| All calves | n=51 | n=54 | n=58 | n=60 | n=55 | n=55 | n=53 | n=56 | n=80 |
| Rate of diarrhoea (%) | 53 | 69 | 69 | 68 | 75 | 15* | 15* | 14* | 39 |
| Duration of diarhoea (days) | 42* | 60 | 48* | 53 | 64 | 55* | 67* | 48* | 84 |
| High intensity of diarrhoea (%) | 31 | 49 | 5 | 5* | 23 | nd | nd | nd | nd |
| Daily gain in body weight (g/d) | 431* | 269 | 378* | 367* | 276 | 292* | 300* | 269* | 185 |
| Rotavirus positive | | | | | | | | | |
| Rate of diarrhea (%) | 80 | 76 | 76 | 78 | 94 | 19* | 44 | 28 | 60 |
| Duration of diarrhoea (days) | 4458 | 44*51 | 6858 | 7248* | 88 | | | | |
| High intensity of diarrhoea (%) | 19 | 50 | 5* | 7* | 50 | nd | nd | nd | nd |
| Daily gain in body weight (g/d) | 462 | 254 | 311* | 300* | 122 | 277 | 238* | 246* | 154 |
| E. coli K99 positive | | | | | | | | | |
| Rate of diarrhea (%) | 31* | 92 | 25 | 33* | 100 | nd | nd | nd | nd |
| Duration of diarrhoea (days) | 60 | 68 | 24 | 54 | 55 | nd | nd | nd | nd |
| High intensity of diarrhoea (%) | 25 | 50 | 0 | 0 | 20 | nd | nd | nd | nd |
| Daily gain in body weight (g/d) | 462* | 262 | 556* | 333 | 189 | nd | nd | nd | nd |

Calves received egg-powder in amounts ranging from 2 to 22 g with (Ab) or without (Co) pathogen-specific egg-antibodies. No egg powder was applied in trial 2 and 3 (Co). In trial 3 the intensity of diarrhoea was not determined, and the calves were too few for results related to Escheria coli K99 to be significant (*p<0.05 vs. control groups of the corresponding trial.). nd = not determined

In the case of the parameters rate, intensity and duration of the diarrhoea, the results of using antibody-powder were often significantly different to the results of using a control-powder or no feed additive. Large differences were recorded in the gain in body-weight. The most important results are summarised in Table 5 and described in more detail by Erhard et al. (1993) and Özpinar et al. (1996).

## Subprotocol 31
## Field Studies of the Prophylactic Use of Specific Egg-Antibodies Against Diarrhoea Due to Infection in Weaned Piglets

### Outline

3 experimental groups:

- Group 1: egg-powder with specific antibodies against E. coli K88, K99, 987P and rotaviruses (antibody-powder)
- Group 2: egg-powder of not immunized hens (control-powder)
- Group 3: without any feed additive (control)

### Procedure

The study on weaned piglets was carried out in a piglet-producing farm in the south of Germany with 70 sows and their 179 weaned piglets.

To limit possible financial harm to the owner, animals whose health was strongly affected were treated additionally with antibiotics.

In the egg-powder the following antibodies were measured:

| | |
|---|---|
| E. coli K88 | (ELISA: at least 470 units) |
| E. coli K99 | (ELISA: at least 100 units) |
| E. coli 987P | (ELISA: at least 860 units) |
| rotaviruses | (neutralization test: at least 512 units) |

1. Suckling of the piglets by the sows in the 7th week of life.
2. Use of the respective powder from the 28th to the 56th day of life. The powder was offered in a feed ration made up of 5% of the mixture and was mixed freely.
3. Diagnosis of pathogens by investigating samples of feces.
4. Evaluation of the parameters: rate and duration of the diarrhoea and gain in body weight additional treatment with antibiotics in the event of severity of symptoms.

## Results

Diarrhoea in suckling piglets was positively affected by the administration of specific egg-powder containing pathogen specific Abs. Whereas diarrhoea occurred in 60% of the piglets in the control group with egg-powder and in 57% in the control group without egg-powder, it occurred in only 17% of the group with antibody-powder.

The efficacy of the antibody-powder is likewise shown by the incidence, duration and intensity of disturbances to the general well-being of piglets in the respective groups. For instance, more than half the piglets in the control groups had to be treated with antibiotics. There were no significant differences in the intensity and duration of diarrhoea in the various groups nor in the gain in weight. The results are shown in Table 6 and described in more detail by Erhard et al. (1996).

# Subprotocol 32
# Field Study of the Therapeutic Use of Specific Egg-Antibodies Against Diarrhoea Due to Infection in Weaned Piglets

## Outline

The therapeutic use of specific egg-antibodies was investigated on newborn piglets in two farms of south Germany which had problems with enterotoxic E. coli (ETEC) infections and diar-

**Table 6.** The prophylactic use of specific egg-antibodies against diarrhoea due to infection in weaned piglets

| Parameters | Control-group (n=58) | Control egg-group (n=61) | Antibody-group (n=60) |
|---|---|---|---|
| Rate of diarrhea (%) | 56.7 | 60.7 | 17.2* |
| Duration of diarrhea (days) | 1.4 (SD 0.7) | 1.5 (SD 0.8) | 1.3 (SD 0.5) |
| High intensity of diarrhea (%) | 62 | 68 | 60 |
| Severity of symptoms (%) | 26.7 | 39.3 | 5.2* |
| Daily gain in body weight (g) | 310 (SD 50) | 280 (SD 30) | 290 (SD 50) |
| Additional treatment (%) | 51.7 | 55.7 | 8.6* |
| Mortality rate (%) | 1.7 | 3.3 | 0.0 |

The animals in the antibody-group or control egg-group were given egg-powder (5% of the ration), respectively with or without pathogen-specific egg-antibodies, whereas the piglets in another control group were given no egg-powder at all (*$p<0.05$ to the control groups, SD = standard deviation).

rhoea. The antibody-titres of egg-powder are according to the trial described under Subprotocol 31 (procedure).

Investigation of 3 groups:

- Group 1: given egg-powder with specific antibodies against E. coli K88, K99, 987P and rotaviruses (antibody-powder)
- Group 2: given egg-powder of not immunised hens (control-powder)
- Group 3: given no feed additive (control)

## Procedure

1. Oral administration of the corresponding egg-powder with a dosage help (3 g per animal per day for 3 days after the start of diarrhoea).
2. Diagnosis of the pathogens.
3. Measurement of the parameters intensity and duration of the diarrhoea.

**Table 7.** Therapeutic use of specific egg-antibodies against diarrhoea due to infection in weaned piglets

| | Control without powder | Control with powder | Control with antibody-powder |
|---|---|---|---|
| Farm 1 | 1.9 | 1.8 | 1.3 |
| Farm 2 | 3.5 | 2.6 | 1.0 |

The numbers indicate the mean duration of diarrhea in days.

## Results

Several samples of feces from piglets suffering from diarrhoea were examined, and hemolytic E. coli with the antigen K88 was found.

On the two farms, altogether 105 new-born piglets got diarrhoea.

On average, the diarrhoea lasted for the number of days as indicated in Table 7. Regarding the intensity of diarrhoea, there were clear differences between the results for piglets treated with antibody-powder and those for the controls. On the whole, diarrhoea occurred more often in the control groups and was more watery and continuous. The results are described in more detail by Kellner et al. (1994).

## Comments

Schmidt et al. (1989) and Wiedemann et al. (1990) were able to show that yolk antibodies, especially in the form of the full egg, are well protected from the processes of digestion. Shimizu et al. (1988) found an acid-stability of purified yolk-antibodies up to pH 4.0. In young animals, whose digestive system is not yet fully functional, orally administered antibodies pass fairly freely through the intestine, so the precondition for the local effect of specific antibodies against diarrhoea due to infections is fulfilled, though little is still known about the concentrations of antibodies needed. A tip about use is given by Yokoyama et al. (1993) in reporting on an infection trial. The specific antibodies proved to be efficacious only if their period of passage through the intestines was long enough.

The advantage of this kind of immune prophylaxis or therapy is that effective action can be taken at any time. It is possible to react immediately and specifically to pathogens, as with a farm-specific vaccine. Care must only be taken to react adequately to a given spectrum of pathogens with a corresponding spectrum of egg-powder antibodies.

This method of local immune-protection helps limit the use of antibiotics to fewer cases and thus limit the development of resistance and imbalance among flora in the intestines. Moreover with the combination of colostral and egg antibodies, the range of antibodies is extended and the efficacy of treatment improved.

In all investigations, irrespective of the kind of animal, of the form in which the antibodies were administered as prophylaxis or therapy, and of the animal's age, the administration of pathogen-specific yolk-antibodies was notably effective (see also Peralta et al. 1994, Ikemori et al. 1996, 1997). An amount of antibodies low in comparison to the concentration of antibodies in the colostrum was enough to thwart pathogens.

*Acknowledgements.* The samples of faeces were examined by Dr. Wittig of the Landesuntersuchungsanstalt für das Gesundheits- und Veterinärwesen (County Health and Veterinary Office) in Saxony as well as by the Tiergesundheitsdienst (Animal Health Service). This studies were supported by Lohmann Animal Health (LAH, Cuxhaven, Germany).

## References

Akita EM, Nakai S (1993) Comparison of four purification methods for the production of immunoglobulins from eggs laid by hens immunized with enterotoxigenic E. coli strain. J. Immunol. Methods 160:207-214

Ambrosius H, Luppa, H. (1987) Immunhistochemie, Grundlagen und Techniken, Akademie-Verlag Berlin, pp. 105-140.

Behn E (1991) Immunodiffusion. In: Friemel H (ed) Immunologische Arbeitsmethoden, 4. Auflage, Gustav Fischer Verlag, Jena, pp. 81-95

Behn I, Hommel U, Oertel M, Hauschildt S (1996) Kinetics of IgY Formation after Immunisation of Hens with different Protein Antigens. Altex 13 Supplement 96:18-21

Benedict AA, Yamaga K(1976) Immunoglobulins and Antibody Production in Avian Species, In: Marchalonis JJ (ed) Comparative Immunology, Blackwell Scient.Publ., Oxford London Edinburgh Australia.

Bjerrum OJ, Schaffner-Nielsen C (1986) Buffer systems and transfer parameters for semidry electroblotting with horizontal apparatus. Electrophoresis 86:315-319

Brumfield HP,Benson H, Pomeroy BS (1961) Procedure for Modified Complement Fixation Test with Turkey, Duck, and Chicken Serum Antibody. Avian Dis 5:270-282

Clarke HGM, Freeman T (1966) A quantitative immuno-electrophorasis method (Laurell electrophoresis). Protides Biol Fluids proc Colloq 14:503-509

Concetti A, Ripani E, Barboni L, Torregiani E, Bombardelli E, Gariboldi P, Venanz FM (1994) Immunorecognition of ring skeleton of taxanes by chicken yolk antibodies. Bilogical Chemistry Hoppe-Seyler 375(6):419-423

Cuelo AC (1983) Immunohistochemistry. IBRO Handbook Series: Methods in the Neurosciences Vol. 3, John Wiley & Sons, Chichester New York Brisbane Toronto Singapore

Danielpour D (1993) Improved sandwich enzyme-linked immunosorbent assays for transforming growth factor $beta_1$. J Immunol Meth 158:17-25

Engvall E (1980) Enzyme Immunoassay ELISA and EMIT. Meth Enzymol 70:419

Erhard MH, Bergmann J, Renner M, Hofmann A, Heinritzi K (1996) Prophylaktische Wirkung von spezifischen Dotterantikörpern bei Escherichia coli K88 (F4)-bedingten Durchfallerkrankungen von Absatzferkeln. (Prophylactic effect of specific egg yolk antibodies in diarrhea of weaned piglets cause by Escherichia coli K88 (F4)). Vet Med A 43:217-223

Erhard MH, Kellner J, Eichelberger J, Lösch U (1993) Neue Möglichkeiten in der oralen Immunprophylaxe des Kalbes - ein Feldversuch mit spezifischen Eiantikörpern (New aspects in oral immunprophylaxis for the prevention of infectious diarrhea of newborn calves - a field study with specific antibodies). Berl Münch Tierärztl Wschr 106:383-387

Erhard MH, Quistorp I von, Schranner I, Jüngling A, Kaspers B, Schmidt P, Kühlmann R (1992) Development of specific enzyme linked immunosorbent antibody assay systems for the detection of chicken immunoglobulins G, M, and A using monoclonal antibodies. Poultry Sci 71:302-310

Erhard MH, Schmidt P, Hofmann A, Bergmann J, Mittermeier P, Kaufmann P, Wiesmüller K-H, Bessler W, Lösch U (1997) The lipopeptid $Pam_3Cys$-Ser-$(Lys)_4$: an alternative adjuvant to Freund's adjuvant for immunisation of chicken to produce egg yolk antibodies. ATLA 25:173-181

Ey PL, Prowse SJ, Jenkin CR (1978) Isolation of pure IgG1, IgG2a and IgG2b immunoglobulins from mouse serum using protein-A sepharose. Immunochem 15:429-457

Fahey JL and McKelvey EM (1965). Quantitative determination of serum immunoglobulins in antibody-agar plates. J Immunol 94:84-91.

Ferencík M. (1993). Handbook of Immunochemistry, Chapman and Hall, London Glasgow New York Tokyo Melbourne Madras. pp. 312-386.

Friemel H (1991) Immunelektrophorese. In: Friemel H (ed) Immunologische Arbeitsmethoden, 4. Auflage, Gustav Fischer Verlag, Jena, pp. 96-107

Goding JW (1976) Conjugation of antibodies with fluorochromes: modifications to the standard methods. J Immunol Meth 13:215-226

Guiterrez JA, Guirriero V (1991) Quantitation of Hsp 70 in tissues using a competitive enzyme-linked immunosorbent assay. J Immunol.Methods 143:81-88

Grabar C, Williams CA (1953) A method permitting the combined study of the electrophoretic and immunochemical properties of a mixture of proteins: application to blood serum. Biophys Acta 10:193-194

Hlinak A, Schrödl W, Witt S, Schade R, Krüger M, (1996) Production of Egg Yolk Antibodies against Human Cell-Associated Antigens Altex 13 Supplement 96:76-79

Ikemori Y, Ohta M, Umeda K, Peralta RC, Kuroki M, Yokoyama H, Kodama Y (1996) Passage of chicken egg yolk antibody treated with hydroxypropyl methylcellulose phtalate in the gastrointestinal tract of calves. J Vet Med Sci 58(4):365-367

Ikemori Y, Ohta M, Umeda K, Icatlo Jr. FC, Kuroki M, Yokoyama H, Kodama Y (1997) Passive protection of neonatal calves against bovine coronavirus-induced diarrhea by administration of egg yolk or colostrum antibody powder. Vet. Microbiol. 58:105-111

Kellner J, Erhard MH, Lösch U (1994) Therapeutischer Einsatz von spezifischen Eiantikörpern bei Säugferkeldurchfall - ein Feldversuch. Tierärztl Umschau 49:31-34

Kowalczyk K, Daiss J, Halpern J, Roth TF (1985) Quantification of maternal-fetal IgG transport in the chicken. Immunol 54:755-762

Kronick MN (1986) The use of phycobiliproteins as fluorescent labels in immunoassay J Immunol Meth 92:1-13

Kyhse-Anderson J (1984) Electroblotting of multiple gels: a simple apparatus without buffer tank for rapid transfer of proteins from polyacrylamide to nitrocellulose. J Biochem Biophys Meth 10:203-209

Laemmli UK (1970) Cleavage of structural proteins during the assembly of the head of bacteriophage T4. Nature 227:655-685

Lanham SM, Godfrey DG (1970) Isolation of salivarian trypanosomes from man and other mammals using DEAE-cellulose. Exp Parasitol 28:521-534

Larsson A, Sjoquist J (1990) Chicken IgY. Utilizing the evolutionary difference. Comp Immun Microbiol Infect Dis 13(4):199-201

Laurell CB (1965) Antigen-antibody crossed electrophoresis. Anal Biochem 10:358-372

Laurell CB (1966) Quantitative estimation of proteins by electrophoresis in agarose gel containing antibodies. Anal Biochem 15:45-52

Lösch U, Schranner I, Wanke R, Jürgens L (1986) The chicken egg, an antibody source. J Vet Med B 98:609-619

Mancini G, Carbonara AO, Heremans JF (1965) Immunochemical quantitation of antigens by simple radial immunodiffusion. Immunochemistry 2:235-254

Marquart WW, Newman JA (1970) A direct complement-fixation test for detection of mycoplasma antibodies in chicken serum. Avian Dis 15, 139-149

Mohr J (1991) Markierung von Antikörpern mit Fluorochromen, In: Friemel H (ed) Immunologische Arbeitsmethoden, Gustav Fischer Verlag, Jena, pp. 135-188

Montes Perez RC, Murcia Meija C, Zarco Quintero L (1994) Production of antibodies against progesterone from the egg yolk of hens and from rabbit blood serum to be used in radioimmunoassay. Veterinaria - Mexico 25(2):117-125

Morino P, Herrera-Marschitz M, Castel M N, Ungerstedt U, Varro A, Dockray G, Hökfelt T (1994). Cholecystokinin in cortico-striatal neurones in the rat: Immunohistochemical studies at the light and microscopical level. Eur J Neurosci 6:681-692

Nakane PK, Kawoi A (1974) Peroxidase antibody. A new method of conjugation. J Histochem Cytochem 22:1084-1090

Ntakarutimana V, Demedts P, van Sande M, Scharpè S (1992) A simple and economical strategy for downstream processing of specific antibodies to human transferrin from egg yolk. J Immunol Methods 153:133-140

O.I.E. (Office International des Epizooties)"Manual of Standards for Diagnostic Tests and Vaccines", Paris 1996

Özpinar H, Erhard MH, Aytug N, Özpinar A, Baklaci C, Karamüptüglu S, Hofmann A, Lösch U (1996) Dose dependent effects of specific egg-yolk antibodies on diarrhoea of newborn calves. Prev Vet Med 27:67-73

Peralta RC, Yokoyama H, Ikemori Y, Kuroki M, Kodama Y (1994) Passive immunisation against experimental salmonellosis in mice by orally administered hen egg-yolk antibodies specific for 14-kDa fimbriae of Salmonella enderitidis. J Med Microbiol 41(1):29-35

Polson A, von Wechmar MB, Fazakerley G (1980) Antibodies to protein from yolk of immunized hens. Immunol Commun 9:495-514

Porstmann T, Porstmann B (991) Enzymimmunoassay. In: Friemel H (ed) Immunologische Arbeitsmethoden, Gustav Fischer Verlag, Jena, pp. 135-188

Reisher JI, Orr HC (1968) Removal of fluorescein isothiocyanate from sephadex after filtration of conjugated proteins. Anal Biochem 26:178-179.

Ressler N (1960) Two-dimensional electrophoresis of serum protein antigens in an antibody containing buffer. Clin Chim Acta 5:795-803

Rieger A, Bürger W, Hiepe W, Schade R (1996) Determination of human serum CRP using a chicken egg yolk antibody (IgY) to avoid interferences with rheumatoid factors. Comparison with mammalian antibodies. ALTEX Suppl. 13:57-61

Rose ME, Orlan E, Buttress N (1974) Immunoglobulin classes in the hens egg: their segregation in yolk and white. Eur J Immunol 4:521-523

Sakamoto N, Takatsuji K, Shiosaka S, Tateishi K, Hashimura E, Miura S, Hamaoka T, Tohyama M (1984). Cholecystokinin-8-like immunoreactivity in the pre- and post-central gyri of the human cerebral cortex. Brain Res 307:77-83

Schade R, Oelßner W, Göhler K, Ott T, Pfister C, Henklein P, Furkert, FJ, Rupprich Ch, Fekete M, Penke B, Jojart, I (1988) Characterization of anti CCK-8 antibodies raised in hens and rabbits by means of several in vitro and in vivo animal models. Biogenic Amines 5:535-550

Schade R, Henklein P, Hlinak A, de Vente J, Steinbusch H (1996) Specificity of chicken (IgY) versus rabbit (IgG) antibodies raised against cholecystokinin octapeptide (CCK). Altex 13 Supplement 96:80-85

Schade R., Pfister C., Halatsch R., Henklein P. (1991) Polyclonal IgY antibodies from chicken egg yolk-an alternative to the production of mammalian IgG type antibodies in rabbits. ATLA 19:403-419.

Schmidt P, Wiedemann V, Kühlmann R, Linckh E, Lösch U (1989) Chicken egg antibodies for prophylaxis and therapy of infectious intestinal diseases. II. In vitro studies on gastric and enteric digestion of egg yolk antibodies specific against pathogenic Escherichia coli strains. J Vet Med B 36:619-628

Schniering A (1995) Untersuchungen zum Koproantigen-Nachweis des Intestinalparasiten Ascaris suum unter Verwendung von aviären vitellinen Antikörpern. Inaugural-Dissertation, Fachbereich Veterinärmedizin der Freien Universität Berlin (Journal Nr. 1837), Germany

Schniering A, Schade R, Hiepe T (1996) Aufbau eines Assays zu Detektion von *Ascaris suum*-Antigenen auf der Basis von IgY. ALTEX Suppl. 13:62-65

Shimizu M, Fitzsimmons RC, Nakai S (1988) Anti- E. coli immunoglobulin Y isolated from egg yolk of immunized chickens as a potential food ingredient. J Food Sci 53:1360-1366

Shimizu M, Nagashima H, Sano K, Hashimoto K, Ozeki M, Tsuda K, Hatta H (1992) Molecular stability of chicken and rabbit immunoglobulin. Biosc Biotech Biochem 56:270-274

Shimizu M, Watanabe A, Tanak A (1995) Detection of high molecular weight mucin-like glcoprotein-A (HMGP-A) of human milk by chicken egg yolk antibody. Biosci Biotech Biochem 59(1):138-139

Shimizu M, Nagashima H, Hashimotot K, Suzuki T (1994) Egg-yolk antibody (IgY) stability in aqueous solution with high sugar concentrations. J Food Sci 59 (4):763

Sofroniew MV (1983) Golgi-like immunoperoxidase staining of neurons producing specific substances or of neurons transporting exogenous tracer proteins. In: Cuello AC (ed) Immunohistochemistry John Wiley and Sons, Chichester, New York, Brisbane Toronto, Singapore, pp. 413-417.

Stolfi RL, Fugmann RA, Jensen JJ, Sigel MM (1971) A C1 Fixation Method for the Measurement of Chicken anti-Viral Antibody. Immunology 20:299-306

Tjissen P (1985) "Laboratory Techniques in Biochemistry and Molecular Biology: Practice and Theory of Enzyme Immunoassays" Elsevier, Amsterdam, New York, Oxford

Towbin H, Gordon J (1984) Immunoblotting and dot immunobinding-current status and outlook. J Immunol Meth 72:313-340

Towbin H, Staehelm T, Gordon J (1979) Electrophoretic transfer of proteins from polyacrylamide gels to nitrocellulose sheets procedure and some applications Proc Natl Acad Sci 76:4259-4354

Van Regenmortel, Burckhard J (1985) Quantitative Microcomplement Fixation tests using Chicken anti-Viral Antibody Extracted from Egg Yolk. J Virol Meth 11:217-223

Vick K (1995) Circadiane Rhythmik von Cholecystokinin und Dopamin im mesolimbischen und nigrostriatalen System der Ratte sowie deren Einfluß auf pharmakologisch induzierte Gehaltsänderungen, Humboldt University, Berlin: Inaugural Dissertation

Wiedemann V, Kühlmann R, Schmidt P, Erhardt W, Lösch U (1990) Chicken egg antibodies for prophylaxis and therapy of infectious intestinal diseases. III. In vivo tenacity test in piglets with artificial jejunal fistula. J Vet Med B 37:283-291

Yokoyama H, Peralta RC, Sendo S, Ikemori Y, Kodama Y (1992) Detection of passage and absorption of chicken egg yolk immunoglobulins in the gastrointestinal tract of pigs by use of enzyme-linked immunosorbent assay and fluorescent antibody testing. Am J Vet Res 54:867-872

## Suppliers

### Egg-powder (Globigen®)

Lohmann Animal Health (Cuxhaven, Germany)

### Labeled secondary anti-hen antibodies are available from several suppliers

BIOGENES
Tel.: 0049-30-65 76 23 96
FAX: 0049-30-65 76 23 97
E-Mail: Biogenes@berlin.sireco.net

BIOZOL
Tel.: 0049-89-319 20 53
FAX: 0049-89-319 36 11
E-Mail: info@biozol.com

CALBIOCHEM
Tel.: 0049-6196-63955 or 56
FAX: 0049-6196-62361
E-Mail: customer.service@calbiochem-novabiochem.de

DIANOVA
Tel.: 0049-40-45 06 70
FAX: 0049-40-45 06 74 90
E-Mail: info@dianova.de

DPC BIERMANN
Tel.: 0049-5221-349 817
FAX: 0049-5221-349-566

GERBU
Tel.: 0049-6223-95 13 0
FAX: 0049-6223-95 13 19
E-Mail: Gerbu@t-online.de

PROMEGA
Tel.: 0049-621-8501 0
FAX: 0049-621-8501 222
E-Mail: de_custserv@de.promega.com

SIGMA
Tel.: 0049-89-6513-1554
FAX: 0049-89-6513-1530
E-Mail: DEorders@eurnotes.sial.com

VECTOR Laboratories
Tel.: 0049-6401-900 77
FAX: 0049-6401-900 78
E-Mail: alexis-d.@alexis-corp.com

# Pros and Cons of Obtaining Antibodies from Chickens' Eggs Rather than from Mammals' Serum

RÜDIGER SCHADE, CHRISTINE SCHWARZKOPF,
and MICHAEL ERHARD

## Cost / benefit relation

The fact that IgY-technology lessens the pain of animals used in experiments should be reason enough for adopting it, but unfortunately the first consideration is generally the cost.

A cage-construction as described under Chapter 2, with four compartments, can accommodate four hens. Such a construction, described by the Swiss National Veterinary Office as adequate for keeping hens in conditions suitable to their nature, costs about 5.500 DM (EBECO). A rabbit-cage (K 99/30 KU, EBECO) of the same make with a floor-area of 5.400 $cm^2$ costs about 4.000 DM and is deemed adequate for keeping not more than two rabbits, depending on their size. In other words, to keep the same number of rabbits, at least two such cages would be needed. Hence an investment of about 5,500 DM would be needed for keeping four hens in comparison with 8,000 DM for keeping two to four rabbits, as long as animal protection is heeded. Since hens count as farmyard-animals, they are cheaper to buy than rabbits, which count as laboratory-animals (see Table 1).

✉ Rüdiger Schade, Universitätsklinikum Charité der Humboldt-Universität, Institut für Pharmakologie und Toxikologie, Dorotheenstr. 94, Berlin, 10117, Germany (*phone* +49-30-20934627; *fax* +49-30-20934608; *e-mail* ruediger.schade@charite.de)
Christine Schwarzkopf, Zentrales Institut des Sanitätsdienstes der Bundeswehr Koblenz, Laborabt. II (Vet.Med), Generaloberst-Beck-Str. 17, Mainz, 55129, Germany
Michael Erhard, Universität Leipzig, Veterinärmedizinische Fakultät, Veterinär-Physiologisches Institut, An den Tierkliniken 7, Leipzig, 04103, Germany

**Table 1.** A Comparison of Outlay for Chicken- and Rabbit-Keeping

| | Laying-Hens (4 of them) | Rabbits (2-4 of them) |
|---|---|---|
| Investment for cage | approx. 5,500 DM | approx. 8,000 DM |
| Price of each animal | approx. 15-20 DM | from 70 DM |
| Cost of maintenance | no general difference | |

According to the German law for animal protection from 1998 (§9, 2) (Bundesgesetzblatt, 1998), rabbits may be used for experiments only if specially bred for the purpose. Hens are not subject to this regulation. These special factors are naturally reflected in the prices.

A comparison between the amounts of antibodies obtainable from hens and from mammals of similar size (rabbits) yields the following balance:

**Table 2.** Amount of Antibodies Obtainable From Rabbit and Chicken

| | IgG (Rabbit) | IgY (Chicken) |
|---|---|---|
| Amount of antibody | approx. 200 mg IgG/bleeding (40 ml blood) | approx. 100 mg IgY/ egg (5-7 eggs/week) |
| Monthly Ab amount | approx. 200 mg IgG | approx. 2,000-2,800 mg IgY |
| Amount of specific Ab | approx. 5 % | 2-10% |

The data are summarised according to a review published in ATLA (Schade et al., 1991) In the literature, the amounts of IgY cited vary from 50 to 250 mg IgY/yolk. This huge discrepancy is due to various factors, such as the marked circadian-rhythm of the IgY content (Schade et al., 1991), the various sizes of yolk, and perhaps racial differences.

As the table reveals, the amount of IgY produced in a month by a laying-hen is about ten times as much as the IgG to be obtained from a rabbit in the same period. Without harming a rabbit, only about 40 ml of blood can be removed per month.

The results of this comparison are summarised in Table 3.

**Table 3.** Comparison of Chicken and Rabbit Concerning Room Allocation and Antibody Output

| | Laying-hen | Rabbit |
|---|---|---|
| Area needed | small animal | small animal |
| Yield of antibodies | big animal | small animal |

This means that the area and associated costs for a laying-hen are no greater than for a rabbit, but the yield of antibodies is such as would otherwise be expected from an animal the size of a sheep or goat and is several times greater than from a rabbit. So the production of antibodies by laying-hens is interesting for keepers whose need of antibodies is rather large and for the first time enables IgY antibodies to be used prophylactically in veterinary medicine and perhaps in future for humans. According to the 3R criteria - reduce, refine, replace - the enormous production of antibodies also means that fewer hens are needed for a specified amount. Hence, given a feasible antibody-titre, the immunisation of a single laying-hen would produce more than enough antibodies to be studied for years in the course of solving a scientific problem. As to the stability of IgY-antibodies, see Chapter 4.

## Special qualities of avian antibodies

Avian antibodies differ from mammalian antibodies in various parameters. Often in the literature these differences are viewed as drawbacks in the use of IgY-antibodies, but this attitude towards the latter is unfair, since their peculiarities have advantages too. In the following table, the differences between IgY- and IgG-antibodies are listed.

**As regards point 1, Table 4**

IgY technology counts as an alternative method in enabling antibodies to be obtained from yolk. If obtaining antibodies from an animal is viewed as being a two-stage process, whose first stage is immunisation and whose second is obtaining antibodies, then in the case of hens one of two invasive and hence distressing procedures is avoided. In effect there is a 50% reduction of animal-distress due normally to immunisation (in the case of mammals).

**Table 4.** Characteristics of Avian IgY in Comparison With Mammalian IgG

| | IgG | IgY |
|---|---|---|
| 1. Ab-sampling | Invasive | non-invasive |
| 2. Protein A/G binding | yes | no |
| 3. Protein L binding | yes* | yes |
| 4. Interference with rheumatoid factor | yes | no |
| 5. Activation of mammalian complement | yes | no |

The data are summarised from a review in ATLA (Schade et al., 1991)

*Except the IgG from sheep, goats and bovines

**As regards point 2/3, Table 4**

Since the Fc-part of the IgY-molecule does not react with the bacterial proteins A or G, IgY cannot be chromatographically purified by having it bind to such proteins, but there are further proteins derived from bacteria with which IgY-antibodies can bind (Serhir et al, 1995). Lately there has been made available a commercial system by which IgY can be purified through affinity chromatography (Fa. Clonetech). The basis of this system is likewise a bacterial cell-wall protein called protein L, which binds various species of immunoglobulins via a K light-chain interaction. The antigen binding is not thereby affected. (DeChateau et al., 1993; Nilson et al., 1992; Akerstrom and Björck, 1988). So the argument that IgY-antibodies cannot be purified in the same way as IgG by affinity chromatography is obsolete. Moreover, since IgY is differently equipped with carbohydrate side-chains (see Chapter 1), purification procedures involving lectines may be feasible.

**As regards point 4, Table 4**

Mammalian antibodies are able to react non-specifically with other mammalian immunoglobulins, for instance with so-called rheumatoid factors. This can lead to false positive results when inflammatory indicator proteins (e.g. C-reactive protein, CRP) in samples of serum from patients with rheumatic ailments are determined by various test-systems (latex-agglutination-test or ELISA) (Rieger et al., 1996). These false positive results are normally avoided in systems based only on mammalian antibodies by blocking them through measures which make the test more elaborate and expensive. The combination with an avian antibody would be an alternative. Since IgY does not cross-react with mammalian immunoglobulins, there is no risk of coming to falsely positive results. According to our own investigations, the percentage of falsely positive results is relatively high if (as shown in Table 5) a test system based upon mammalian antibodies is used without additional steps to avoid non-specific reaction with rheumatoid factors.

**Table 5.** CRP-Latex Test (Sheep Anti-Human CRP Antibody)

| | CRP positive | CRP negative | |
|---|---|---|---|
| rheumatoid factor positive | 39 | 6 | Σ45 |
| rheumatoid factor negative | 11 | 8 | Σ19 |
| | Σ50 | Σ14 | Σ64 |

Modified from Rieger et al. (1996)

**Table 6.** CRP-Latex Test (Chicken Anti-Human CRP Antibody)

| | CRP positive | CRP negative | |
|---|---|---|---|
| rheumatoid factor positive | 22 | 23 | Σ45 |
| rheumatoid factor negative | 10 | 9 | Σ19 |
| | Σ32 | Σ32 | Σ64 |

Modified from Rieger et al. (1996)

**As regards point 5, Table 4**

It is sometimes said to be a drawback of avian antibodies that they are unable to activate complement, but the claim is too general. Avian antibodies can indeed activate the avian complement system though not the mammalian one. This can be very useful, if activation of the complement system is not desired. If mammalian antibodies are taken as the sole criterion for evaluating the antibodies of other species, prejudices may arise which on the one hand are hard to overcome and on the other are not factually justified. Moreover, even within the class of mammals there are considerable differences in structural and functional properties of antibodies (Ambrosius and Rudolph, 1990).

In our opinion, avian antibodies are neither better nor worse than mammalian ones but mainly different. They have properties to be exploited in most of the various test-systems, especially in combination with mammalian antibodies.

Besides the already mentioned differences between mammalian and avian antibodies, the hens' immune system has properties favourable for producing antibodies against highly preserved mammalian antigens. It has been shown in many ways that laying hens, as opposed to mammals, react by producing specific antibodies when immunised with such antigens (Goueli et al., 1990; Gassman et al., 1993; Lee et al., 1995). The cause of this immune response may lie in the phylogenetic difference between the classes of avians and mammals. Moreover there are structural differences between IgY and IgG, which enable various epitopes (sequence- and structure-determinants) to be recognised (e.g. Gerl et al., 1996). So, even from the scientific point of view, the immunising of laying hens is a viable alternative to the immunising of rabbits.

## References


Akerstrom B, Björck L (1988) Protein L - an immunoglobulin light chain-binding bacterial protein: characterization of binding and physicochemical properties. J Biol Chem 264:19740-19746

Ambrosius R, Rudolph W (1990) Grundriß der Immunbiologie, Second Edition, Gustav Fischer, Stuttgart New York, pp. 54-71

Bundesgesetzblatt Jahrgang 1998 Teil I Nr. (1998) Bekanntmachung der Neufassung des Tierschutzgesetzes vom 25. Mai 1998 30, ausgegeben zu Bonn am 29. Mai 1998

DeChateau M, Nilson BHK, Erntell M, Myhre E, Magnusson CGM, Akerström B, Björck L (1993) On the interaction between protein L and immunoglobulins of various mammalian species. Scand J Immunol 37:399-405

Gassmann M, Thömmes P, Weiser T, Hübscher U (1990) Efficient production of chicken egg yolk antibodies against a conserved mammalian protein, FASEB J 4:2528-2532

Gerl M, Steinert C, Quint M, Schade R, Günzler V (1996) Immunisation of chickens with the aminoterminal propeptide of bovine procollagen type III. Altex, Suppl. 13:51-56

Goueli SA, Hanten JD, Ahmed K (1990) Polyclonal antibodies against rat liver cytosolic casein kinase II (CK-2) cross-react with CK-2 from other tissues and nuclear form (PK-N2) of the enzyme. Biochem. Int. 21:685-694

Lee YI, Surzycki SS, Lee YI (1995) Production of egg yolk antibody (IgY) against human placental DANN-dependent RNA polymerase II. J Biochem Mol Biol 28:27-32

Nilson BHK, Lögdberg L, Kastern W, Björck L, Akerström B (1993) Purification of antibodies using protein L-binding framework structures in the light chain variable domain. J Immunol Methods 164:33-40

Rieger A, Bürger W, Hiepe W, Schade R (1996) Determination of human serum CRP using a chicken egg yolk antibody (IgY) to avoid interferences with rheumatoid factors. Comparison with mammalian antibodies. Altex Suppl. 13:57-61

Schade R, Pfister C, Halatsch R, Henklein P (1991) Polyclonal IgY antibodies from chicken egg yolk - an alternative to the production of mammalian IgG type antibodies in rabbits. ATLA 19:403-419

Serhir B, Dubreuil D, Higgins R, Jacques M (1995) Purification and characterization of a 52 kilodalton immunoglobulin G-binding protein from Streptococcus suis capsular type 2. J Bacteriol 177:3830-3836

# IgY-Antibodies, Future Prospects

Rüdiger Schade

In the past two decades, biomedical research has been confronted with a growing public interest in the welfare of animals used in research. The resulting discussion has encouraged the search for ways to refine, reduce or even replace animal experiments. This also applies to the production of mono- and polyclonal antibodies in animals. These antibodies are important tools in bio-medical research, e.g. they are essential components of diagnostic methods for quantitative and qualitative determinations of a wide variety of biological molecules. Polyclonal antibodies are produced in rabbits and other rodents like mice, rats and guinea-pigs or in farm animals such as horses, sheep and goats. The production of polyclonal antibodies involves two stressful procedures, firstly the immunisation procedure with several injections and secondly the harvesting of the antibodies which involves bleeding of the animals.

For more than a century it has been known (Klemperer, 1893; see Figure 1) that in immunised hens the concentration of antibodies is almost the same as in serum. However, there is conflicting evidence in the literature, e.g. Larsson et al. reported in 1993 that the concentration of the IgY antibody was higher in yolk than blood concentrations. Although several classes of immunoglobulin are found in the serum of hens, only IgY is found in substantial amounts in yolk (see ECVAM report, Schade et al., 1996). Chicken IgY has sometimes been called IgG, since this Ig isotype exhibits properties quite similar to mammalian IgG. Moreover, IgY is sometimes called IgA, since the

Rüdiger Schade, Universitätsklinikum Charité der Humboldt-Universität, Institut für Pharmakologie und Toxikologie, Dorotheenstr. 94, Berlin, 10117, Germany (*phone* +49-30-20934627; *fax* +49-30-20934608; *e-mail* ruediger.schade@charite.de)

first immunological studies in chicken were first evaluated from a phylogenetical point of view. If hens are used for the production of antibodies, in comparison to mammals the painful step of blood sampling is replaced by antibody-extraction from egg yolk, which is abundant in comparison to blood. Thus the production of polyclonal antibodies in chicken is not only an improvement but also a valuable alternative in terms of the three Rs principle of Russel and Burch (1959) - refine, reduce, replace - for the humane conduction of animal experiments. We have termed the new procedure producing polyclonal antibodies the IgY-technology, since the procedure consists of several distinct methodologies, as e.g. accommodation of hens, laying behaviour, immunisation protocols, extraction procedures and, in particular, the specific molecular structure of IgY, biochemical properties of IgY favouring a typical pattern of application in biomedical research. By now the term 'IgY-technology' is well accepted both in the scientific literature and for commercial products containing avian IgY antibodies.

Although it is obvious that hens offer advantages as a source of antibodies (see Chapter 6), they are not yet widely used for this purpose. This seems to be due to a variety of factors, as e.g. the habits of investigators, lack of experience and information. Generally two points seem to be of particular concern, first the keeping of hens (see Chapter 2), and secondly the harvesting of antibodies by extraction from egg yolk. Neither claim is up-to-date, since systems for keeping hens and IgY extraction procedures are well established, including IgY-affinity-chromatography on IMSORB (JsV) or KAPTIV-GY™.

Despite the described prejudices, interest in IgY antibodies has been increasing, when judging either from the number of publications on IgY or from the number of secondary-antibodies, extraction-kits and diagnostics that are commercially available. Unfortunately, a real breakthrough has not yet occurred although avian antibodies have become known to potential users, which will eventually favour their application. Thus an essential step forward has been made during the last decade.

Today there are two major directions in the application of the IgY technology:

# ARCHIV
FÜR
# EXPERIMENTELLE PATHOLOGIE
UND
# PHARMAKOLOGIE

HERAUSGEGEBEN VON

PROF. R. BOEHM IN LEIPZIG, PROF. O. BOLLINGER IN MÜNCHEN, PROF. E. BOSTRÖM IN GIESSEN, PROF. C. GAEHTGENS IN GIESSEN, PROF. E. HARNACK IN HALLE, PROF. F. A. HOFFMANN IN LEIPZIG, PROF. F. HOFMEISTER IN PRAG, PROF. M. JAFFÉ IN KÖNIGSBERG, PROF. E. KLEBS IN CARLSRUHE, PROF. PH. KNOLL IN PRAG, PROF. E. KÜLZ IN MARBURG, PROF. TH. LANGHANS IN BERN, PROF. L. LICHTHEIM IN KÖNIGSBERG, PROF. W. MARMÉ IN GÖTTINGEN, PROF. HANS MEYER IN MARBURG, PROF. B. NAUNYN IN STRASSBURG, PROF. M. v. NENCKI IN ST. PETERSBURG, PROF. E. NEUMANN IN KÖNIGSBERG, PROF. F. PENZOLDT IN ERLANGEN, PROF. H. QUINCKE IN KIEL, PROF. F. v. RECKLINGHAUSEN IN STRASSBURG, PROF. F. RIEGEL IN GIESSEN, DR. L. RIESS IN BERLIN, PROF. O. SCHMIEDEBERG IN STRASSBURG, PROF. JUL. SCHREIBER IN KÖNIGSBERG, PROF. H. SCHULZ IN GREIFSWALD, PROF. R. THOMA IN DORPAT, PROF. C. WEIGERT IN FRANKFURT A. M.

REDIGIRT VON

Dr. B. NAUNYN UND Dr. O. SCHMIEDEBERG

PROFESSOR DER MEDICINISCHEN KLINIK — PROFESSOR DER PHARMAKOLOGIE

IN STRASSBURG I. E.

EINUNDDREISSIGSTER BAND.

MIT 25 ABBILDUNGEN IM TEXT UND 1 TAFEL.

LEIPZIG,
VERLAG VON F. C. W. VOGEL.
1893.

XV.

Aus dem Laboratorium der medicinischen Klinik zu Strassburg i. E.

## Ueber natürliche Immunität und ihre Verwerthung für die Immunisirungstherapie.

Von

Dr. Felix Klemperer,
Assistent der Klinik.

Fig. 1. The newspaper and the title of the article which more than a hundred years ago proved the presence of neutralising antibodies in the yolk of eggs from immunised hens.

- Application of the IgY technology driven by progress in immunology.
  It seems realistic to anticipate that the IgY-technology will be accepted as one of several methods for producing polyclonal antibodies since technical problems have been overcome and the method can easily be applied. However, since scientists are fairly conservative and usually stick to methods that have worked for them it is unrealistic to assume that polyclonal avian antibodies will generally replace mammalian ones in bio-medical research and diagnostic procedures. Thus it is more likely that IgY-antibodies will be used in immunology to study specific problems and they will, in addition, cover a small but constant part of the general production of polyclonal antibodies. The application of avian antibodies in human medicine for therapeutic purposes is currently beyond our imagination.

- Application of the IgY technology driven by legal regulations in favour of animal protection.
  Within the next decade the implementation of EU Directive 86/609/EEC for the protection of animals used for experiments and other purposes will stimulate efforts to increasingly apply alternatives to experimental animals. Consequently, hens rather than rabbits will be immunised, and in more general terms scientists may even be asked to produce antibodies without using animals. In the long run the application of molecular genetics may allow to produce antibodies without using animals at the same quality and price as traditional polyclonal antibodies. Once this stage has been reached, classical polyclonal and monoclonal antibodies produced in animals will not be required any more. However, the properties of antibodies produced by gene-technology will first have to be shown in daily use. The first production of monoclonal antibodies was welcomed with high expectations, all of which were not met when the new technology was applied. In fact, polyclonal antibodies are still in use; and due to their immunological properties, they are quite often superior to monoclonals. Hence the production of monoclonal antibodies by gene-technology will not immediately stop the immunisation of animals, and, in contrast to the expectations of the animal rights movement and leg-

islators, in the near future antibodies from gene-technology may not be the perfect alternative to the production of polyclonal antibodies in animals.

For the reasons explained we are convinced that under the present legal environment the immunisation of hens should be encouraged in all research funded by the government. Consequently, the scope of argumentation put forward by scientists who cannot be convinced to use the IgY technology (see above) would be curtailed or eliminated by legal measures. This scenario would, however, presuppose adequate accommodation for the hens.

When summarising, there seems to be growing interest in using avian IgY antibodies. This interest has not very often convinced scientists to establish the new technology in their laboratories, but they are now using it more often than a few years ago. Although the production of IgY-antibodies reduces the suffering of laboratory animals, this consideration does not sufficiently convince scientists to adopt the technology, as they are in the first place interested in the advancement of science. This general element and the availability of the IgY technology are the most crucial factors influencing the decision of scientists whether or not to apply the new technology. A more widespread implementation of the IgY technology will depend on the following factors: first of all, feasibility, which has been established, secondly scientific potential, and finally legislation favouring the reduction of numbers and suffering of laboratory animals.

## References


Klemperer F (1893) Ueber natürliche Immunität und ihre Verwerthung für die Immunisirungstherapie. Arch. Exptl. Pathol. Pharmakol. 31:356-382

Larsson A, Balöw R-M, Lindahl TL, Forsberg P-O (1993) Chicken antibodies: Taking advantage of evolution - A Review. Poultry Sciences 72:1807-1812

Russell WMS, Burch RL (1959) The principles of human experimental technique. Methuen, London

Schade R, Staak C, Hendriksen C, Erhard M, Hugl H, Koch G, Larsson A, Pollmann W, van Regenmortel M, Rijke E, Spielmann H, Steinbusch H, Straughan D (1996) The production of avian (egg yolk) antibodies. IgY. The report and recommendation of ECVAM workshop 21. ATLA 24:925-934.

# Appendix

RÜDIGER SCHADE

## Materials

Buffers

**Coating buffer (ELISA) (carbonate-buffer pH 9.6; 0.1 M, other molarities are possible e.g. 0.05 M)**

- 8.37 g $Na_2CO_3$ x 10 $H_2O$
- 6.00 g $NaHCO_3$
- Dissolve in 1000 ml dist. water

**PBS (phosphate buffered saline, pH 7.2-7.4)**

- 8.0 g NaCl
- 2.9 g $Na_2HPO_4$ x 12 $H_2O$
- 0.2 g $KH_2PO_4$
- 0.2 g KCl
- 0.2 g $NaN_3$
- Dissolve in 1000 ml dist. water

**TBS (tris-buffered saline, pH 7.3)**

- 12.11 g Tris

Rüdiger Schade, Universitätsklinikum Charité der Humboldt-Universität, Institut für Pharmakologie und Toxikologie, Dorotheenstr. 94, Berlin, 10117, Germany (*phone* +49-30-20934627; *fax* +49-30-20934608; *e-mail* ruediger.schade@charite.de)

- 2.05 g NaCl
- 0.75 g Glycine
- 0.2 g $NaN_3$
- Dissolve in 800 ml dist. water
- Adjust the pH and fill up with dist. water to 1000 ml

**RIA-buffer (pH 7.2 - 7.4)**

- 7.01 g NaCl
- 0.36 g $KH_2PO_4$
- 1.89 g $Na_2HPO_4$ x 12 $H_2O$
- 1.00 g $NaN_3$
- Dissolve in 1000 ml dist. water
- Add 10 % (v/v) Haemaccel 35 (Behring Werke AG)

**Blot-buffer (pH to 7.2 - 7.4)**

- 0.59 g $KH_2PO_4$
- 5.81 g $Na_2HPO_4$ x 12 $H_2O$
- 0.40 g KCl
- 17.72 g NaCl
- Dissolve in 2000 ml dist. Water.
- Add 2 ml Triton X

**Electrophoresis buffer**

- Part A
  - 112.4 g Glycin
  - 90.4 g Tris
  - Dissolve in 2000 ml dist. Water.
- Part B
  - 26 g Veronal-sodium
  - 4.2 g Veronal (dissolve in boiled water)
  - Dissolve in 2000 ml dist. water.

**Note:** Before use mix 1 part buffer A with 1 part buffer B and dilute with 2 parts dist. water (pH 8.5 - 8.7, ionic strength 0.04).

## Staining of immunoprecipitates

**Staining solution**

- 5 g Coomassie brilliant blue G250
- 450 ml ethanol (96 %)
- 450 ml dist. water
- 100 ml glacial acetic acid
  Dissolve the staining in the acid ethanol and leave at room temperature until the next day. After filtration the solution is ready for use.

Destain the gels by using the same solution but without Coomassie brillant blue. The destaining solution can be regenerated by being passed through a carbon filter.

## TMB

**Substrate solutions for ELISA**

- A. Stock solution
  - Dissolve tetramethyl benzidine in dimethyl sulphoxide at a concentration of 6 mg/ml and store about 10 ml in a dark bottle.

**Note:** Warning: TMB is poisonous. DMSO is less poisonous

- B. TMB-Puffer (0.1 mol/l sodium acetate citrate buffer; pH 5.0)
  - 8.20 g sodium acetate
  - 3.15 g citric acid
  - Add 1.0 l aqua bidest
- C. Substrate solution for application

**Per plate:**
- 332 µl stock solution
- 10 ml TMB buffer (warm before in an incubator and add slowly to avoid precipitation)
- 3 µl 30 %ige $H_2O_2$ (add to a beaker, mix carefully and use at once, since the resulting solution is light- sensitive and effective for not more than 4 min)

- D. Stop-reagent (1-molar sulphuric acid; pH 1)
  - Put 472 ml aqua bidest into a flask,
  - add 28 ml 96% $H_2SO_4$ via a measuring retort slowly, as heat is produced fast.

**Note:** Warning: $H_2SO_4$ is corrosive.

- E. Use
  - 1. Start a stop watch,
  - 2. Add 100 µl of the substrate-solution in each well.
  - 3. Leave the plate in the dark.
  - 4. After exactly 10 min reaction, add 50 µl stop reagent to each well.
  - 5. Measure immediately at 450 nm in the ELISA-reader, the reference filter being 620 nm.

**Note:** For health safety reasons the use of commercial products is recommended (ready for use solution: Seramun-Blau [Seramun Diagnostica GmbH], TMB tablets [SIGMA]

**OPD**

- A. Substrate solution (per plate)
  - 11 ml Phosphate-citrate buffer pH 5.0
    - 24 mM citric acid
    - 50 mM $NH_2HPO_4$ x $H_2O$
  - 5.5 µl 30 % $H_2O_2$
  - 3 mg 1.2-phenylendiamine
- B. Stop-reagent
  - 1 molar sulphuric acid as with TBM
- C. Use
  - As with TBM, but read the plate at 492 nm.

**Note:** For health safety reasons the use of commercial products is recommended (OPD tablets [SIGMA])

**ABTS**

- A. Substrate solution (Phosphate-citrate buffer pH 5.0)
  - Buffer 1:
    - citric acid 4.2 g
    - dist. water add 200 ml
  - Buffer 2:
    - $Na_2HPO_4$ x 12 $H_2O$ 7.16 g
    - dist. water add 200 ml
- Phosphate-citrate buffer is prepared by mixing equal amounts of buffer 1 and buffer 2.

- Dissolve 40 mg of ABTS (2,2'-Azino-bis[3-ethylbenzthiazoline-6-sulfonic acid]) in 100 ml phosphate-citrate buffer. Store at $4^0$C in the dark. Just before use add 100 µl of 1:40 diluted $H_2O_2$ (30%) to 10 ml ABTS.
- B. Stop-reagent
  - The reaction may be stopped with 1% SDS or 10% liqid soap (ordinary dish washing agent).
- C. Use
  - As with TBM but read the plate at 405 nm.

**Note:** For health safety reasons the use of commercial products is recommended (ABTS tablets [SIGMA])

**Ammonium sulphate solutions**
- For producing ammonium sulphate solutions see Table 1 (Holtzhauer, 1995).

## Specifics of IgY-technology

IgY technology has specific features which potential users often feel are more complicated than in fact they are. In the following, information is offered about practical details which may encourage potential users to produce antibodies in hens.

### Typical titre-developments in laying-hens

As regards the production of antibodies in laying-hens, the user will be interested in how the titre develops after immunisation. As in the case of mammals, titre-development seems to depend on the immunogenity of the antigen. Basically the development takes one of two forms. The first is more or less saltatory in the sense that after a boost the development of antibodies rises for about 10 days then reaches a constant level lasting for about another 10 days before falling. After a second boost, the same happens again (see Figures 2, 4).We have seen such developments after immunisation with peptide-KLH conjugates for instance (see also the review from Schade et al., 1992). The other titre-development tallies more closely with usual expectations: after the boost the antibody-development rises then remains for

**Table 1.** Producing ammonium sulphate solutions. Table 1a Ammonium sulphate saturation at 25°C ; Table 1b Ammonium sulphate saturation at 0°C

Table 1a Ammonium sulphate saturation at 25°C

| F | 10 | 20 | 30 | 40 | 45 | 50 | 55 | 60 | 65 | 70 | 75 | 80 | 90 | 100 |
|---|---|---|---|---|---|---|---|---|---|---|---|---|---|---|
| I | | | | | | | | | | | | | | |
| 0 | 56 | 114 | 176 | 243 | 277 | 313 | 351 | 390 | 430 | 472 | 516 | 561 | 662 | 767 |
| 5 | 27 | 85 | 140 | 210 | 248 | 286 | 318 | 355 | 400 | 441 | 480 | 522 | 618 | 715 |
| 10 | | 57 | 118 | 183 | 216 | 251 | 288 | 326 | 365 | 406 | 449 | 494 | 592 | 694 |
| 20 | | | 59 | 123 | 155 | 189 | 225 | 262 | 300 | 340 | 382 | 424 | 520 | 619 |
| 30 | | | | 62 | 94 | 127 | 162 | 198 | 235 | 273 | 314 | 356 | 449 | 546 |
| 40 | | | | | 31 | 63 | 97 | 132 | 168 | 205 | 245 | 283 | 375 | 469 |
| 45 | | | | | | 32 | 65 | 99 | 134 | 171 | 210 | 250 | 339 | 431 |
| 50 | | | | | | | 33 | 66 | 101 | 137 | 176 | 214 | 302 | 392 |
| 60 | | | | | | | | | 34 | 69 | 105 | 143 | 227 | 314 |
| 70 | | | | | | | | | | | 35 | 72 | 153 | 237 |
| 75 | | | | | | | | | | | | 36 | 115 | 198 |
| 80 | | | | | | | | | | | | | 77 | 157 |
| 90 | | | | | | | | | | | | | | 77 |

Table 1b Ammonium sulphate saturation at 0°C

| F | 10 | 20 | 30 | 40 | 45 | 50 | 55 | 60 | 65 | 70 | 75 | 80 | 90 | 100 |
|---|---|---|---|---|---|---|---|---|---|---|---|---|---|---|
| I | | | | | | | | | | | | | | |
| 0 | 55 | 107 | 166 | 229 | 262 | 295 | 331 | 366 | 404 | 442 | 483 | 523 | 611 | 707 |
| 5 | 26 | 80 | 139 | 200 | 232 | 266 | 300 | 336 | 373 | 411 | 450 | 491 | 578 | 671 |
| 10 | | 54 | 111 | 171 | 203 | 236 | 270 | 305 | 342 | 379 | 418 | 458 | 544 | 636 |
| 20 | | | 56 | 84 | 115 | 145 | 177 | 210 | 244 | 280 | 316 | 392 | 476 | 565 |
| 30 | | | | 57 | 87 | 119 | 150 | 184 | 217 | 253 | 289 | 328 | 408 | 495 |
| 40 | | | | | 29 | 59 | 90 | 122 | 155 | 190 | 225 | 262 | 340 | 424 |
| 45 | | | | | | 29 | 60 | 91 | 125 | 158 | 193 | 229 | 306 | 388 |
| 50 | | | | | | | 30 | 61 | 93 | 127 | 161 | 197 | 272 | 353 |
| 60 | | | | | | | | | 31 | 63 | 96 | 131 | 204 | 283 |
| 70 | | | | | | | | | | | 32 | 66 | 136 | 212 |
| 75 | | | | | | | | | | | | 32 | 102 | 176 |
| 80 | | | | | | | | | | | | | 68 | 141 |
| 90 | | | | | | | | | | | | | | 71 |

I = initial saturation in %

F = desired final saturation in %

The numbers show the amount of ammonium sulphate (g) which has to be added to a 1000 ml solution at saturation I to produce saturation F.

a longer time at the level reached (see Figures 1, 3 ,5 ,6). Of interest in this respect is the fact that hens kept under SPF conditions do not basically achieve any different titre-development to hens kept conventionally (see Figures 5, 6; Behn et al. unpublished). Such antibody-developments are often observed after immunisation with proteins. The titre-developments shown here reveal that the immunisation of hens leads to titre-developments comparable to those in rabbits.

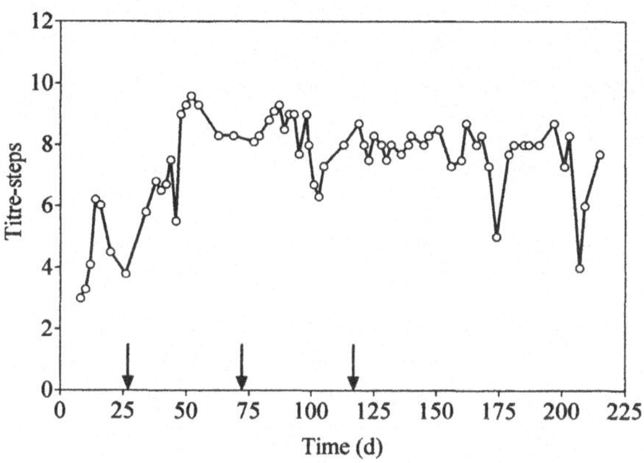


**Fig. 1.** Antibody titre development after immunising hens (n = 5) with rabbit IgG. The arrows point to the times of boost immunisations. Since the hens were kept as a group, it is impossible to say which eggs are from which hens, so the curve is partly irregular. Otherwise, in the long run, the antibody-titre fluctuates around titre-level 8. The titre levels are $1{:}100x2^n$, with n ranging from 0 to 9 i.e. the values range from 1:100 to 1:51,200.

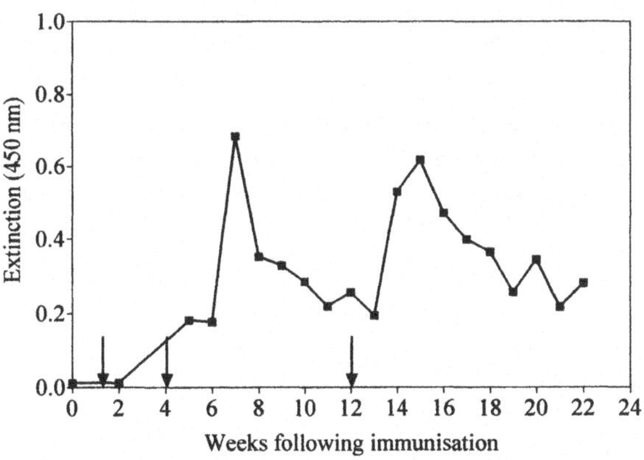


**Fig. 2.** Average antibody-titre development after immunising hens (n = 6) with antigens of *Cryptosporidium* spec. (FCA, Mittermeier, 1995). Use was made of a yolk-extract diluted 1:20,000. Each time after a boost, there was a rise in the titre. The arrows indicate boost immunisations.

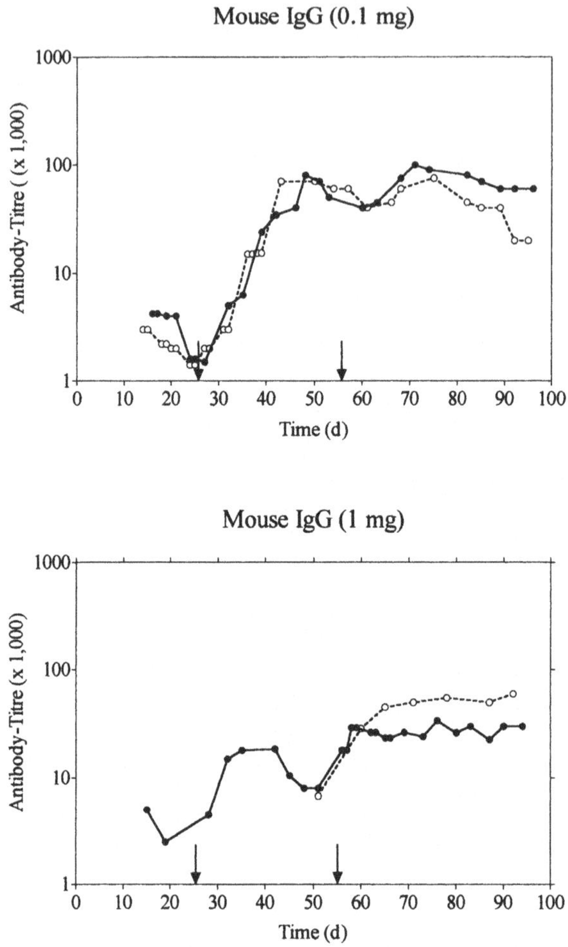


**Fig. 3.** Antibody-titre development after immunising hens (n = 2) with mouse IgG. The values on the y-axis on the left are to be multiplied by 1.000. The antibody-titre development seems to depend on the amount of antigen used. In the case of immunisation with 0.1 mg IgG, high titres are reached already between the 40th and the 50th day after the 1st boost. The arrows mark the times of the boosts. From Behn et al. (1996).

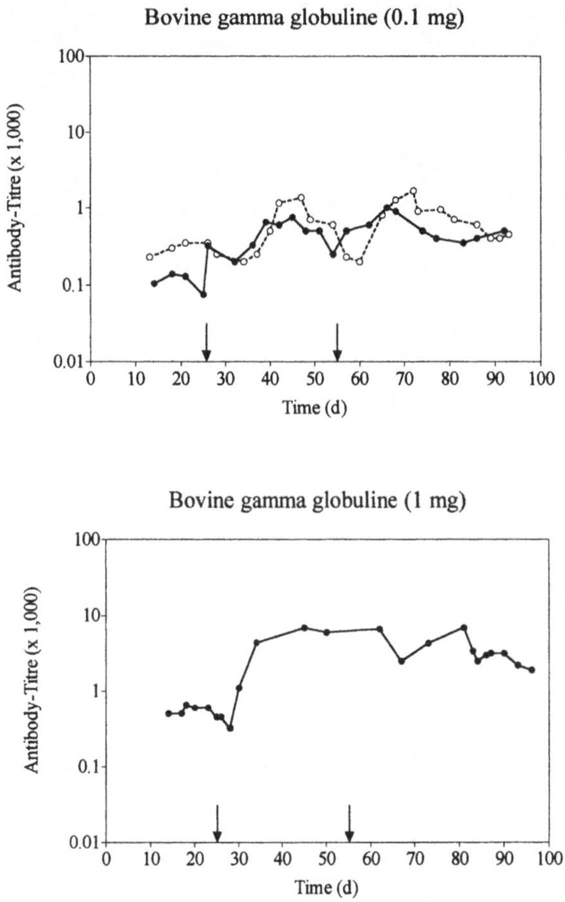


**Fig. 4.** Antibody-titre development after immunising hens (n = 2) with bovine gammaglobulin. In this case there are clearly higher titres after using 1 mg BGG. Otherwise the results were as in Figure 3.

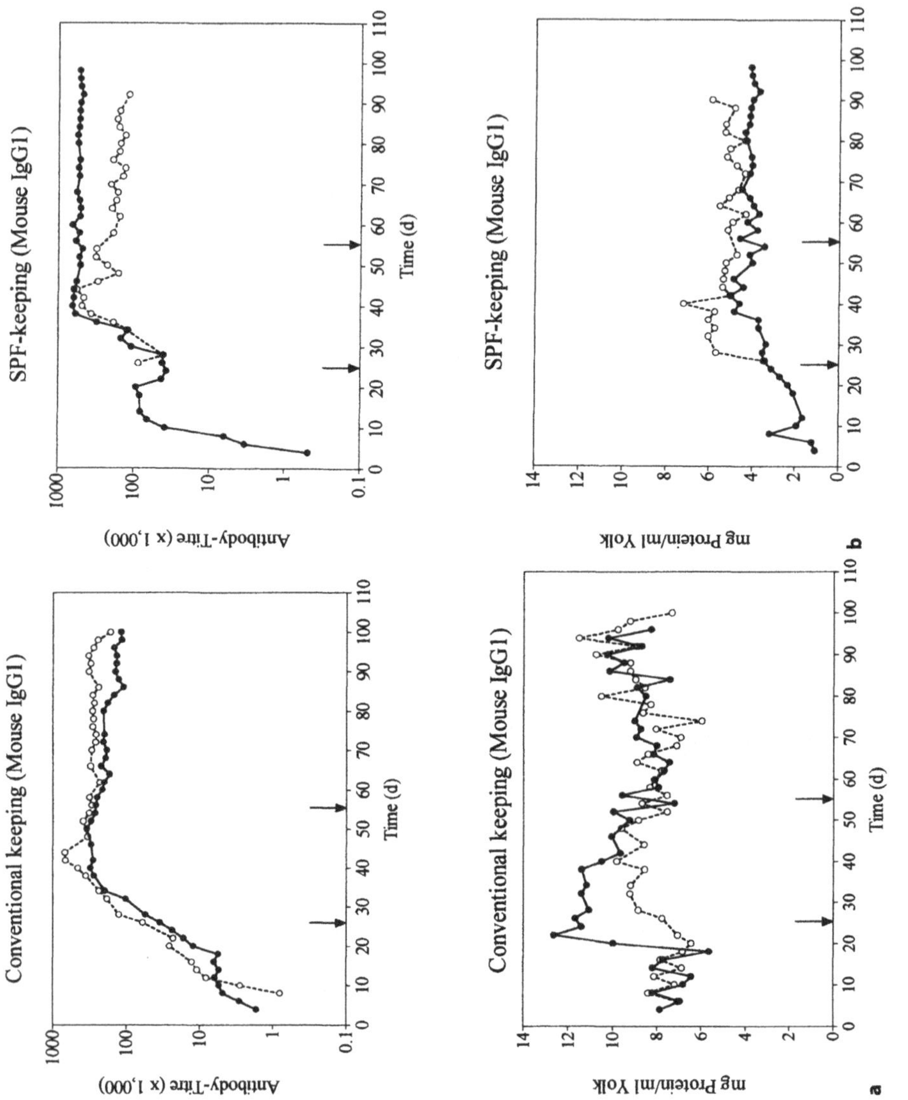


**Fig. 5.** A comparison of the antibody-titre development of each of 2 hens, one being kept conventionally and the other under SPF-conditions, after immunisation with mouse IgG1 (the boost immunisation was performed at the same time as in Figure 3. The hens both belonged to the race 'white leghorn'. **a** Conventional keeping. In the upper part is the antibody-titre development, and in the lower part the time-dependent change of the yolk-protein content. Note the evident seasonal fluctuations of the protein content, which is the same in both titres. Otherwise the results were as in Figure 3). **b** SPF keeping. As in Figure 5a. Note that the titre-developments in both cases are similar. The yolk-protein content is markedly less under SPF conditions and rises clearly after immunisation. No rhythmic changes of protein concentration are evident. Otherwise the results were as in Figure 3 (Behn et al. unpublished).

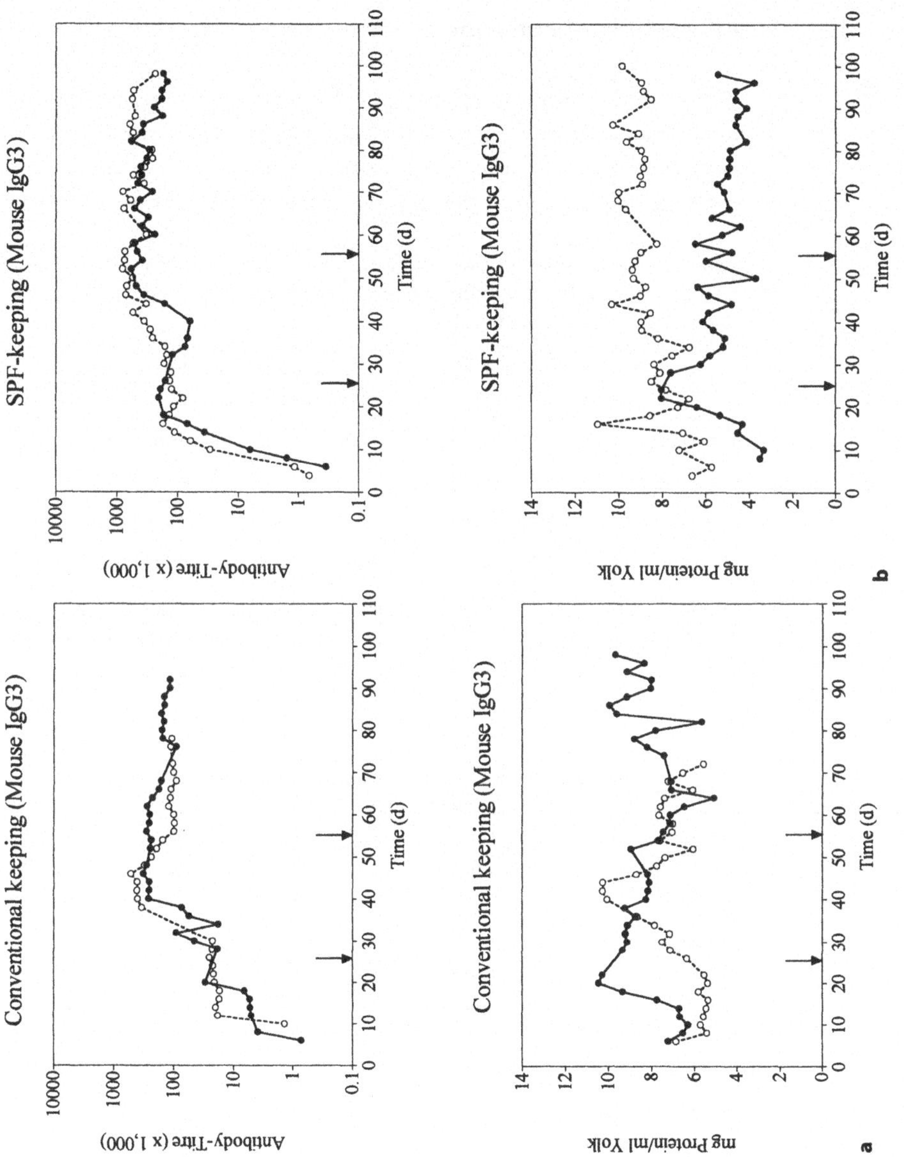


**Fig. 6.** Comparison of the antibody-titre development of each of 2 hens, one being kept conventionally and the other under SPF-condition, after immunisation with mouse IgG3 as in Figure 5 (the boost immunisation was performed at the same time as in Figure 3). **a** Conventional keeping. **b** SPF keeping. In this case too, there is a notable difference in the development of the yolk-protein concentration (Behn et al. unpublished).

## Information about eggs

Since IgY-technology relies basically on eggs, here are some relevant data. They relate on the one hand to the constitution of eggs (see Table 2) and on the other to the increase in volume with increasing age of the hen (see Figure 7; Mittermeier, 1995).

**Table 2.** Constitution of the egg of an adult laying-hen of the race White Leghorn (n=10, the value for egg-white has been calculated, the values in brackets are from Männer 1987)

| | Weight in g | SD (Standard Deviation) | % of whole |
|---|---|---|---|
| Whole egg | 65.5 (58.1) | 4.3 | 100 |
| Egg white | 39.8 (33.8) | | 60.8 (58.1) |
| Egg yolk | 18.5 (18.5) | 2.1 | 28.2 (31.8) |
| Shell | 7.2 (5.8) | 0.6 | 11.0 (10.1) |

Lyophilisation of the egg (without shell) leaves on average 14.1 g egg-powder per egg, i.e. 23.6 % of the fluid whole egg. (From Mittermeier 1995)

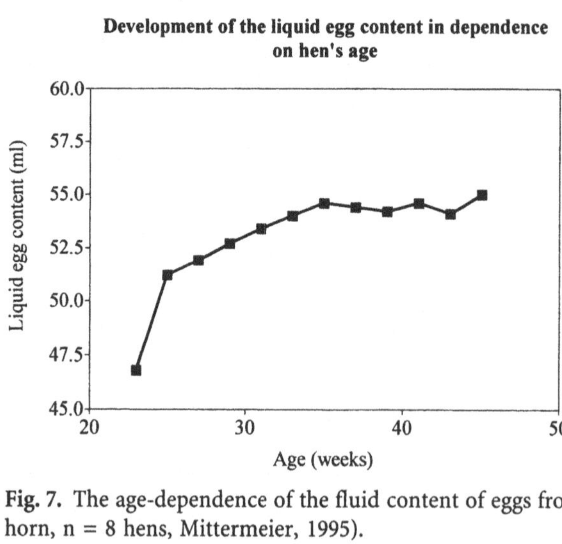


**Fig. 7.** The age-dependence of the fluid content of eggs from hens (white leghorn, n = 8 hens, Mittermeier, 1995).

## Information about the optimal temperature for antibody-binding in ELISA

The body temperature of mammals ranges from 30 to 41°C, and of birds from 38 to 44°C (Hesse and Doflein, 1935). Normally EIA-plates are incubated at 37°C, to approximate to the average body-temperature of mammals, but the body-temperature of hens is well above this (41.5°C), so some experiments were performed at higher temperatures. Figure 8 shows that the optimum of extinction is in fact about 42°C, but the difference to the rate at 37°C is slight (about 0.1), so we have not changed the protocol. An incubation time of 60 min is sufficient to obtain a nearly maximal extinction (Figure 10).

## Information about time-dependence of IgY-binding (RIA)

The binding-kinetics of an antibody in RIA can vary a great deal. Sometimes 24 hours are enough to reach maximal binding, but not always (see Strohbach 1994). In our experience good results can be achieved with an incubation time of about 24 hours (see Chapter 5), though the specific binding in certain cases takes 72 hours for completion and is then about 20 % more (see Figure 9).

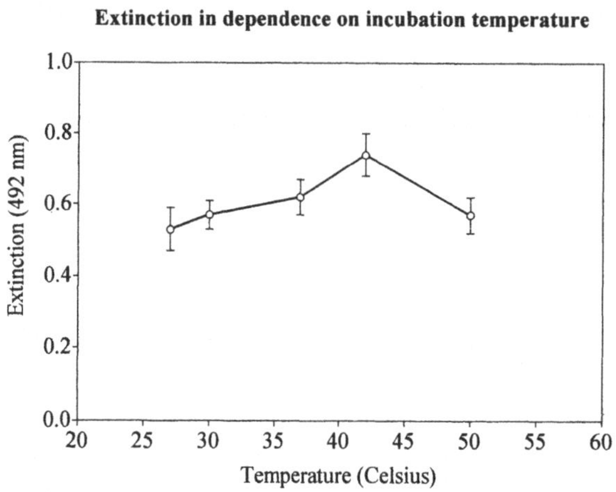


**Fig. 8.** The temperature-dependence of the binding of an IgY antibody to a tallying antigen (n = 10 repeats SD). The optimal temperature turns out to be about 42°C, which is more or less the body-temperature of laying hens.

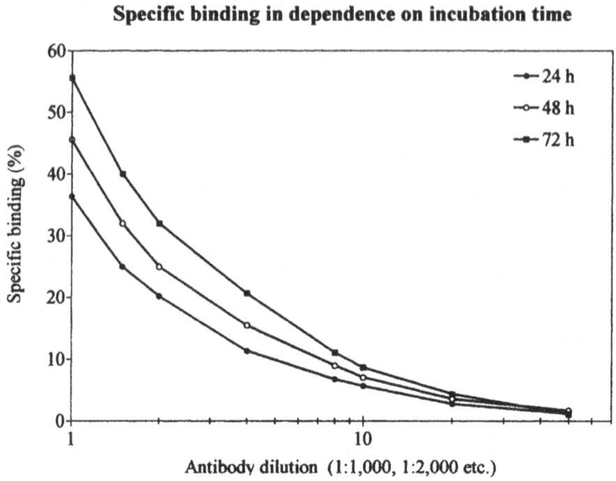


**Fig. 9.** The time-dependence of the binding of an IgY antibody to a corresponding antigen (CCK8) in a radio immunoassay. The greatest binding is reached after 72 h, which is about 20% more than after 24 h.

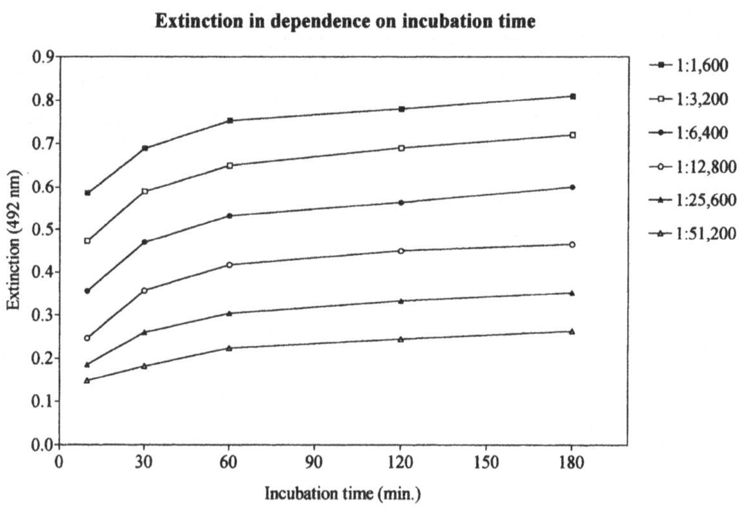


**Fig. 10.** The time-dependence of the binding of an IgY antibody to a tallying antigen in an ELISA (according to Figure 8). There is a practicable incubation-time of about 60 min.

### Information about seasonal fluctuations in the yolk IgY-content

Information about yolk IgY-content varies a great deal in the literature (see Chapter 4). This may be the cause of the controversy about whether or not yolk IgY-content is greater than serum IgY-content. According to our research, yolk IgY is subject to infradian changes of up to 100 % (see Figure 11). Behn et al. have arrived at similar results (unpublished; see Figures 5a, 6a). Extraordinarily interesting and not yet described is the observation that although the yolk IgY-content of hens kept conventionally is subject to seasonal fluctuations, the same seems not to be true of hens kept under SPF-conditions (see Figures 5b, 6b). The meaning of this finding is unclear.

## Procedure

### Detailed protocol for IgY-extraction by means of PEG

**Equipment**

- stirrer
- cooling centrifuge

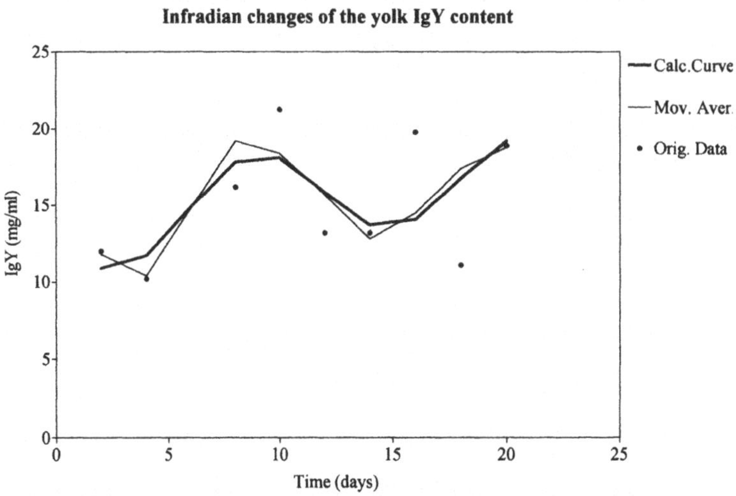


**Fig. 11.** The time-dependent change in yolk IgY-content. The original values were analysed according to the procedure of the moving average (thin line) or the curve was smoothed out by the PHARMAFIT program (Mattes et al., 1991).

- glass equipment
- paper filter
- pipettes
- photometer
- centrifuge test-tubes with screw-stoppers

**Reagents, solutions**

- phosphate-buffer (egg-buffer) pH 7.5
  - 50 ml 0.2 M $Na_2HPO_4$
  - 5 ml 0.2 M $NaH_2PO_4$
  - 100 ml 1.0 M NaCl
  - 10 ml 1.0 % $NaN_3$ (toxic)
  - distilled water (add to 1000 ml and adjust the pH)
- PEG 6000 (pulverised by a mortar)

**Procedure**

1. Crack the egg open carefully and, without harming the yolk, separate it from the white egg and put it in a measuring cylinder, containing 10 ml egg-buffer already.
2. Record the yolk volume, which is the total volume minus 10 ml buffer.
3. Transfer the materials to a 100 ml glass beaker.
4. Stir and mix the yolk with a glass rod and add egg-buffer (2 x yolk-volume minus 10 ml).
5. Put the beaker onto a magnet stirrer to be stirred moderately for about 2 min.
6. Add 0.035 g PEG 6000/ml total volume (see Table 3, that corresponds to a PEG concentration of 3.5 %, see also Chapter 4).
7. Have it stirred mildly for about 2 min.
8. Transfer the solution to centrifuge test-tubes.
9. Centrifuge at 11.000 rpm for 10 min at 4°C (5.000 - 10.000 x g).
10. Filter the supernatant through filter-paper in a funnel into a 100 ml measuring cylinder, measure the volume of the filtrate and transfer it to a 100 ml glass beaker.

11. Add 0.085 g PEG in portions /ml filtrate (see Table 3, that corresponds to a final PEG concentration of 12 %, see also Chapter 4).
12. Have it stirred mildly for 10 min.
13. Transfer the solution into centrifuge test-tubes and centrifuge it as above.
14. Discard the supernatant.
15. Dissolve the sediment in egg-buffer (1 x yolk-volume) and transfer it into the beaker already used.
16. Add 0.12 g PEG in portions /ml yolk-volume (that corresponds to a PEG concentration of 12 %, see also Chapter 4).
17. Have it stirred mildly for 10 min.
18. Centrifuge as above.
19. Discard the supernatant.
20. Dissolve the sediment in egg-buffer (1/2 x yolk-volume).
21. Determine the proteins at 280 nm.
22. Dialyse the IgY-extract against egg-buffer.
23. Aliquote the extract and store it at -20 °C.

An overview of general publications on IgY-Antibodies is given in Table 4.

**Table 3.** Calculated values according to the PEG method

| Filtrat (ml) | 0.085 g PEG/ ml Filtrate | Egg volume (ml) | | | 0.035 g PEG/ ml | 0.12 g PEG/ ml |
|---|---|---|---|---|---|---|
| | | 1/2 x | 1x | 2x | | |
| 19 | 1.62 | 4 | 8 | 16 | 0.84 | 0.96 |
| 20 | 1.7 | 4.5 | 9 | 18 | 0.95 | 1.08 |
| 21 | 1.79 | 5 | 10 | 20 | 1.05 | 1.20 |
| 22 | 1.87 | 5.25 | 10.5 | 21 | 1.10 | 1.26 |
| 23 | 1.96 | 5.5 | 11 | 22 | 1.16 | 1.32 |
| 24 | 2.04 | 5.75 | 11.5 | 23 | 1.20 | 1.38 |
| 25 | 2.13 | 6 | 12 | 24 | 1.26 | 1.44 |
| 26 | 2.21 | 6.25 | 12.5 | 25 | 1.31 | 1.50 |
| 27 | 2.3 | 6.5 | 13 | 26 | 1.37 | 1.56 |
| 28 | 2.38 | 6.75 | 13.5 | 27 | 1.42 | 1.62 |
| 29 | 2.47 | 7 | 14 | 28 | 1.47 | 1.68 |
| 30 | 2.55 | 7.25 | 14.5 | 29 | 1.52 | 1.74 |
| 31 | 2.64 | 7.5 | 15 | 30 | 1.58 | 1.80 |
| 32 | 2.72 | 7.75 | 15.5 | 31 | 1.63 | 1.86 |
| 33 | 2.81 | 8 | 16 | 32 | 1.68 | 1.92 |
| 34 | 2.89 | 8.25 | 16.5 | 33 | 1.73 | 1.98 |
| 35 | 2.98 | 8.5 | 17 | 34 | 1.79 | 2.04 |
| 36 | 3.06 | 8.75 | 17.5 | 35 | 1.84 | 2.10 |
| 37 | 3.15 | 9 | 18 | 36 | 1.89 | 2.16 |
| 38 | 3.23 | 9.25 | 18.5 | 37 | 1.94 | 2.22 |
| 39 | 3.32 | 9.5 | 19 | 38 | 2.00 | 2.28 |
| 40 | 3.4 | 9.75 | 19.5 | 39 | 2.05 | 2.34 |
| 41 | 3.49 | 10 | 20 | 40 | 2.10 | 2.40 |
| 42 | 3.57 | 10.25 | 20.5 | 41 | 2.15 | 2.46 |
| 43 | 3.66 | 10.5 | 21 | 42 | 2.21 | 2.52 |
| 44 | 3.74 | 10.75 | 21.5 | 43 | 2.26 | 2.58 |
| 45 | 3.83 | 11 | 22 | 44 | 2.31 | 2.64 |
| 46 | 3.91 | 11.25 | 22.5 | 45 | 2.36 | 2.70 |

**Table 4.** Overview of general publications on "IgY - technology" (The table makes no claim to be complete)

| General Publications | | |
|---|---|---|
| Camenisch G et al. | 1999 | FASEB 13, 81-88 |
| Schade R et al. | 1997 | ATLA 13, Suppl. 96, 179 - 183 |
| Fischer M, Hlinak A | 1996 | ALTEX 13, Suppl. 96, 179 - 183 |
| Fischer M et al. | 1996 | Tierärztl. Praxis, 24, 411 - 418 |
| Hlinak A et al. | 1996 | Tierärztl. Umschau 51, 402 - 408 |
| Schade R, Hlinak A | 1996 | ALTEX 13, Suppl. 96, 5-9 |
| Svendsen L et al. | 1995 | Lab. Anim. Sci. 45, 89 - 93 |
| Schade R et al. | 1994 | ALTEX 11, 75 - 83 |
| Schade R et al. | 1994 | ALTEX 2, 75-84 |
| Larsson A et al. | 1993 | Poultry Sci. 72, 1807 - 1812 |
| Erhardt MH et al. | 1992 | Poultry Sci. 71, 302 - 310 |
| Larsson A et al. | 1991 | Clin. Chem. 37, 411 - 414 |
| Schade R et al. | 1991 | ATLA 19, 403 - 419 |
| Gassmann M et al. | 1990 | Schweiz. Arch. Tierheilk. 132, 289 - 294 |
| Larsson A, Sjöquist J | 1990 | Comp. Immun. Microbiol. Infect. Dis. 13, 199 - 201 |
| Polson A | 1990 | Immun. Invest. 19, 253 - 258 |
| Wallmann J et al. | 1990 | J. Vet. Med. B, 37, 317 - 320 |
| Lösch U et al. | 1986 | J. Vet. Med. B, 33, 609 - 619 |
| Bade H, Stegemann H | 1984 | J. Immunol. Meth. 72, 421 -426 |
| Jensenius JC et al. | 1981 | J. Immunol. Meth. 46, 63 - 68 |

## Applications

Table 5 gives an overview of publications on diagnostic and therapeutic applications of specific egg yolk antibodies (IgY). This table does not at all claim to be complete.

**Table 5.** Diagnostic and therapeutic applications of specific egg yolk antibodies (IgY)

| | | | |
|---|---|---|---|
| A Diagnostic applications | | | |
| a) Egg yolk antibodies against virus antigens | | | |
| Cymbidium mosaic virus | Vejaratpimol R et al. | 1998 | J. Ferm. Biotechn. 86, 65 - 71 |
| Newcastle disease virus | Gervelmeyer A et al. | 1998 | Revue Elev. Med. vet. Pays trop. 51, 1 , 5 - 9 |
| Infectious bronchitis virus | Gervelmeyer A et al. | 1998 | Revue Elev. Med.vet. Pays trop. 51, 1 , 5 - 9 |
| Gumboro disease virus | Gervelmeyer A et al. | 1998 | Revue Elev.Med. vet. Pays trop. 51, 1 , 5 - 9 |
| Bovine leukaemia virus gp 51, p 24 | Schade R et al. | 1997 | ATLA 25, 555 - 586 |
| Infectious bronchitis virus (IBV) | Shafiq MK et al. | 1997 | Vet. Med.J., Giza. 45, 3, 273 |
| Coronavirus, bovine | Ikemori Y et al. | 1997 | Vet. Microbiol. 58, 105 - 111 |
| Infectious bronchitis virus (IBV) | Staak C | 1996 | ALTEX 13 (Suppl.), 73 - 75 |
| Rabies virus | Staak C | 1996 | ALTEX 13 (Suppl.), 73 - 75 |
| Newcastle disease virus (NDV) | Staak C | 1996 | ALTEX 13 (Suppl.), 73 - 75 |
| Gumboro disease virus (IBDV) | Staak C | 1996 | ALTEX 13 (Suppl.), 73 - 75 |
| Newcastle disease virus (NDV) | Afzal H et al. | 1996 | Pakistan Vet. J. 16, 4 ,196 -199 |
| Hepatitis B surface antigen (HBsAg) | Gross M, Speck J | 1996 | Dtsch. tierärztl. Wschr. 103, 417 - 422 |
| Potato viruses (PVA, PVS, PVM, PVX, PVY) | Gross M, Speck J | 1996 | Dtsch.tierärztl. Wschr. 103, 417 - 422 |
| Rotavirus | Kuroki M et al. | 1994 | Arch. Virol. 138, 143-148 |
| Rotavirus, bovin | Kuroki M et al. | 1993 | Vet. Microbiol. 37, 135-146 |
| Potato virus Y | Polson A | 1993 | Prep. Biochem. 23, 273 - 284 |
| Influenza virusdifferent strains | Cuceanu N et al. | 1991 | Roum. Arch. Microbiol. Immunol. 50, 215-222 |

**Table 5.** Continuous

| | | | |
|---|---|---|---|
| Rotavirus, human | Ebina T et al. | 1990 | Microbiol. Immunol. 34, 7, 617 - 629 |
| Adenoviruses , hexon | Khilko SN et al. | 1989 | Mol. Gen. Microbiol. Virol. 9, 43-48 |
| Plant viruses | Rowland GF et al. | 1989 | J. Virol. Meth. 25, 259 - 269 |
| Coronavirus, bovine | Czerny CP, Eichhorn W | 1989 | Vet. Microbiol. 20, 111 - 122 |
| Canine distemper virus | Schmidt P et al. | 1989 | J. Vet. Med. B 36 , 661 - 668 |
| Enteroviruses | Graewskaja NA et al. | 1988 | Mh. Vet. Med. 43, 677-680 |
| Yellow fever virus | Graewskaja NA et al.. | 1988 | Mh. Vet. Med. 43, 677-680 |
| Hepatitis virus A | Graewskaja NA et al. | 1988 | Mh. Vet. Med. 43, 677-680 |
| Rotavirus, human | Yolken RH et al. | 1988 | Pediatrics 81, 291-295 |
| Influenza virus | Döller PC et al. | 1987 | Med. Microbiol. Virol. 176, 27-35 |
| Respiratory syncytial virus (RSV) | Zrein M et al. | 1986 | Arch. Virol. 90, 197 - 206 |
| Rotavirus | Regenmortel MHV van, Burkhard J | 1985 | J. Virol. Meth. 11, 217 - 223 |
| Plant viruses | Regenmortel MHV van, Burkhard J | 1985 | J. Virol. Meth. 11, 217 - 223 |
| Newcastle disease virus (NDV) | Piela TH et al. | 1984 | Avian Dis. 28, 4 , 877 |
| Infectious bronchitis virus (IBV) | Piela TH et al. | 1984 | Avian Dis. 28, 4 , 877 |
| Rotavirus, murine | Bartz CR et al. | 1980 | J. Infect. Dis. 142, 439 - 441 |
| Plant viruses | Bar-Joseph M, Malkinson M | 1980 | J. Virol. Meth. 1 , 179 - 183 |
| b) egg yolk antibodies against bacterial antigens | | | |
| Salmonella enterica | Terzolo et al. | 1998 | Rev. Arg. Microbiol. 30, 84-92 |
| Clostridium botulinumtype C | Rocke TE et al. | 1998 | J. Wildlife Dis. 34, 744 - 751 |
| Lipopolysaccarid | Sasse M et al. | 1997 | Berl. Münch. Tierärztl. Wschr. 111, 121 - 126 |
| Lipopolysaccarid | Sasse M, Hlinak A | 1997 | Berl. Münch. Tierärztl. Wschr. 111, 127 - 133 |

**Table 5.** Continuous

| | | | |
|---|---|---|---|
| lipid - A | Sasse M et al. | 1997 | Berl. Münch. Tierärztl. Wschr. 111, 121 - 126 |
| Lipid - A | Sasse M, Hlinak A | 1997 | Berl. Münch. Tierärztl. Wschr. 111, 127 - 133 |
| Clostridium botulinum, neurotoxin A, B, E | Doellgast GJ et al. | 1997 | J.Clin.Microbiol. 35, 3, 578 - 583 |
| Salmonella enteritidis | Gast RK et al. | 1997 | Poultry Science 76, 798 - 801 |
| Salmonella enteritidis | Mine Y | 1997 | J. Agric. Food Chem. 45 , 3723 - 3727 |
| Lipopolysaccharide (LPS), Salmonella enteritidis | Mine Y | 1997 | J. Agric. Food Chem. 45 , 3723 - 3727 |
| Bordetella pertussis | Schade R et al. | 1997 | ATLA 25, 555 - 586 |
| polysaccharide, bacterial (alginate) | Gross M, Speck J | 1996 | Dtsch. tierärztl. Wschr. 103, 417 - 422 |
| Bordetella bronchiseptica | Hlinak A et al. | 1996 | ALTEX 13 (Suppl.), 70 - 72 |
| Aeromonas salmonicida ssp. | Denzin N, Henrion E | 1996 | Bull. Europ. Ass. Fish Pathol. 16, 47 - 50 |
| Chlamydia psittaci | Zahn I et al. | 1995 | J. Vet. Med. B 42 , 266 - 276 |
| Chlamydia trachomatis | Zahn I et al. | 1995 | J. Vet. Med. B 42 , 266 - 276 |
| E.coli | Tanimoto T et al. | 1995 | Nipp. Shok. Kag. Kag. Kaishi 42, 5, 225 - 229 |
| Salmonella enteritidisfimbrien antigen | Peralta RC al. | 1994 | J. Med. Microbiol. 41, 29-35 |
| E.coli | Akita EM, Nakai S | 1993 | J. Immunol. Meth. 160, 207 - 214 |
| E. coli, fimbrien antigens | Ikemori Y et al. | 1993 | Poult. Sci. 72, 2361-2365 |
| E.coli | Yokoyama H et al. | 1993 | Am. J. Vet. Med. 54 , 6 , 867 |
| Rickettsia conorii | Kalvatchev Z et al. | 1993 | Acta Virol. 37 , 184 - 186 |
| E.coli | Akita EM, Nakai S | 1992 | J. Food. Sci. 57, 629 - 634 |
| E. coli , K99 | Ikemori Y et al. | 1992 | Am. J. Vet. Res. 53, 11 , 2005 |
| staphylococcal protein A (SpA) | Godfrey MA et al. | 1992 | J. Immunol. Meth. 149, 21 - 27 |
| Escherichia (E.) coli | Wiedemann V et al. | 1991 | J. Vet. Med. B 38, 283-291 |
| Escherichia (E.) colidifferent strains | Jüngling A et al. | 1991 | J. Vet. Med. B 38 , 373 - 381 |

**Table 5.** Continuous

| | | | |
|---|---|---|---|
| Escherichia (E.) coli | Wiedemann V et al. | 1991 | J. Vet. Med. B 38 , 283 - 291 |
| Chlamydiae spp. | Strautmann Kunz U et al. | 1991 | J. Vet. Med. B 38, 292 - 298 |
| Corynebacterium sepedonicum | Underberg HA, Sander E | 1991 | Z. Pflanzenkd. Pflanzenschutz 98, 188 - 196 |
| S.typhimurium | Schiemann DA, Montgomery AL | 1991 | Vet. Microbiol. 27, 295 - 308 |
| Escherichia (E.) coli | Wiedemann V et al. | 1990 | J. Vet. Med. B 37 , 163 - 172 |
| Salmonella spp. S.typhimurium, S. pullorum | Wallmann J et al. | 1990 | J. Vet. Med. B 37 , 317 - 320 |
| Bradyrhizobium bacterium | Gillette WK et al. | 1990 | Fed. Am. Soc. Exp. Biol. J. 4, 2070 |
| Escherichia (E.) coli | Schmidt P et al. | 1989 | J. Vet. Med. B 36 , 619 - 628 |
| Bacteroides amylophilus | Ricke SC et al. | 1988 | Appl. Environ. Microbiol. 54, 596-599 |
| Bacteroides rumenocola, subsp. brevis | Ricke SC et al. | 1988 | Appl. Environ. Microbiol. 54, 596-599 |
| Selenomonas ruminatium, subsp. rum.D | Ricke SC et al. | 1988 | Appl. Env. Microbiol. 56, 2795 - 2800 |
| Acholeplasma laidlawii | Hakkarainen K et al. | 1987 | Isr. J. Med. Sci. 23, 539 |
| Brucella abortus | Lösch U et al. | 1986 | J. Vet Med. B 33, 609 - 619 |
| Mycoplasma gallisepticum | PielaTH et al. | 1984 | Avian Dis. 28, 877-833 |
| Legionella pneumophila | Rodgers FG | 1982 | H. Med. Microbiol. 15, 181-188 |
| Mycoplasma gallisepticum | Mohammed HO et al. | 1986 | Avian Dis. 30, 2 , 389 |
| Mycoplasma synovia | Mohammed HO et al. | 1986 | Avian Dis. 30, 2 , 389 |
| c) egg yolk antibodies against antigens of parasites and other pathogens | | | |
| T-2 mycotoxin | Kierek-Jasczuk D et al. | 1997 | J. Food Protection 60, 3, 321 - 327 |
| Ascaris suum | Schniering A et al. | 1996 | ALTEX 13 (Suppl.), 62 - 65 |
| Cryptosporidi... | Schmidt P et al. | 1996 | ALTEX 13 (Suppl.), 30 - 34 |
| Coccidien | Lillihoj HS et al. | 1993 | Avian Dis. 37, 731-740 |

**Table 5.** Continuous

| | | | |
|---|---|---|---|
| Toxoplasma gondii | Hassl A, Aspöck H | 1988 | J. Immunol. Meth. 110, 225 - 228 |
| Sarcocystis gigantea | Hiepe F et al. | 1988 | Mh. Vet. Med. 43, 680 - 682 |
| d) egg yolk antibodies against proteins and peptides | | | |
| cholecystokinin octapeptide (CCK-8) | Schade R et al. | 1997 | ATLA 25, 555 - 586 |
| alpha-antiplasmin, human | Lee SW et al. | 1997 | Prep. Biochem. Biotechn. 27, 4, 227 - 237 |
| Cathepsin D | Fortgens PH et al. | 1997 | Immunopharm. 36 , 305 - 311 |
| Procollagen type III, bovine, propeptid | Gerl M et al. | 1996 | ALTEX 13 (Suppl.), 51 - 56 |
| Troponin I, human, cardiac | Doth M et al. | 1996 | ALTEX 13 (Suppl.), 54 - 50 |
| Somatotropin, recombinant, bovine | Hofmann A et al. | 1996 | ALTEX 13 (Suppl.), 26 - 29 |
| C-reactive protein (CRP) | Rieger A et al. | 1996 | ALTEX 13 (Suppl.), 57 - 61 |
| CD - 14 | Hlinak A et al. | 1996 | ALTEX 13 (Suppl.), 76 - 79 |
| Spiramycin | Albrecht U | 1996 | Milchwiss. 51, 4 , 209 - 212 |
| Gamma globuline, bovine, dinitrophenylated | Behn I et al. | 1996 | ALTEX 13 (Suppl.), 18 - 21 |
| p 23, cellulare | Hlinak A et al. | 1996 | ALTEX 13 (Suppl.), 76 - 79 |
| IgG, mouse | Hommel U et al. | 1996 | ALTEX 13 (Suppl.), 40 - 44 |
| IgM, mouse | Hommel U et al. | 1996 | ALTEX 13 (Suppl.), 40 - 44 |
| IgG, subclasses, murine | Erhardt MH et al. | 1996 | ALTEX 13 (Suppl.), 66 - 69 |
| IgG, rabbit | Schwarzkopf Ch, Thiele B | 1996 | ALTEX 13 (Suppl.), 35 - 39 |
| IgG, fox | Schwarzkopf Ch, Thiele B | 1996 | ALTEX 13 (Suppl.), 22 - 25 |
| IgM, mouse | Behn I et al. | 1996 | ALTEX 13 (Suppl.), 18 - 21 |
| IgG, mouse | Behn I et al. | 1996 | ALTEX 13 (Suppl.), 18 - 21 |
| albumin, serum, human | Schmidt P et al. | 1996 | ALTEX 13 (Suppl.), 30 - 34 |
| IgG, dog | Staak C | 1996 | ALTEX 13 (Suppl.), 73 - 75 |
| IgG, horse | Staak C | 1996 | ALTEX 13 (Suppl.), 73 - 75 |
| IgG, warthog | Staak C | 1996 | ALTEX 13 (Suppl.), 73 - 75 |

**Table 5.** Continuous

| | | | |
|---|---|---|---|
| Interleukin-6 (IL-6) | Wooley JA, Landon J | 1995 | J. Immunol. Meth. 178, 253 - 265 |
| DAB III | Concetti A et al. | 1994 | Biol. Chem. Hoppe-Seyler 375, 419 - 423 |
| Beta-casekinin -10 | Meisel A | 1994 | Biol. Chem. Hoppe-Seyler 375, 401 -405 |
| Soybean glycinin A | Meisel H | 1993 | Int. Dairy J. 3 , 149 - 161 |
| Parathyroid hormone-related protein | Rosol TJ et al. | 1993 | Vet. Immunol. Immunopathol. 35, 321 - 337 |
| Growth hormone | Schmidt P et al. | 1993 | J. Histochem. Cytochem. 41, 1441 - 1446 |
| Prolactin | Schmidt P et al. | 1993 | J. Histochem. Cytochem. 41, 1441 - 1446 |
| Alpha-amidating enzyme | Sturmer AM et al. | 1992 | J. Immunol. Meth. 146, 105 -110 |
| Transferrin, human | Ntakarutimana V et al. | 1992 | J. Immunol. Meth. 153, 133 - 140 |
| von Willebrand factor | Toti F et al. | 1992 | Haemostasis 22, 32 - 40 |
| Glycoprotein GP IIb-IIIa | Toti F et al. | 1992 | Haemostasis 22, 32 - 40 |
| Sperm, human | Polson A | 1991 | Immunol. Invest. 20, 451 - 459 |
| Insulin, human | Lee K et al. | 1991 | Agric. Biol. Chem. 55, 2141 - 2143 |
| Hsp 70, bovine tissue | Gutierrez A, Guerriero V | 1991 | J. Immunol. Meth. 143, 81 - 88 |
| Blood group B, human | Hoomi O et al. | 1991 | Anim. Sci. Technol. 62, 11, 1055 - 1061 |
| Lactoferrin | Meisel H | 1990 | Milchwirtsch. 45 , 510 - 512 |
| Dihydroxyvitamin D, 1,25 | Bauwens RM et al. | 1988 | Clin. Chem. 34, 2153 - 2154 |
| IgG, human | Hiepe F et al. | 1988 | Mh. Vet. Med. 43, 680 - 682 |
| IgD, human | Hiepe F et al. | 1988 | Mh. Vet. Med. 43, 680 - 682 |
| Dihydroxyvitamin D, 1,25 | Bauwens RM et al. | 1987 | Clin. Chem. Acta 170, 37 - 44 |
| Soybean proteins | Eichler D, Rubach K | 1986 | Z. Lebensm. Unters. Forsch. 183 , 8 - 11 |
| Kallikrein, human plasma | Burger D et al. | 1985 | Thromb. Res. 40, 283 - 288 |
| Insulin receptor, alpha subunit | Song CS et al. | 1985 | J. Immunol. 135, 3354 - 3359 |
| Parathyroid hormon, human | Vieira JGH et al. | 1984 | J. Immunoass. 5, 121 - 129 |

**Table 5.** Continuous

| | | | |
|---|---|---|---|
| Thymus RNA polymerase II | Carroll SB, Stollar BD | 1983 | J. Biol. Chem. 258, 24 - 26 |
| Prostaglandin | Fertel R et al. | 1981 | Biochem. Biophys. Res. Comm. 102, 1028 - 1033 |
| Venom | Polson A et al. | 1980 | Immunol. Commun. 9, 5, 495 - 514 |
| B Therapeutic applications of egg yolk antibodies (IgY) | | | |
| Pathogens in newborn calves (infectious diarrhea) | Erhardt MH et al. | 1997 | Arch. Tierernährung 50, 4, 369 -380 |
| P 110 protein human stomach cancer MGC-803 cells | Yang J et al. | 1997 | Chin. J. Biotechn. 13, 85 - 90 |
| Streptococcus mutans dental plaque formation | Hatta H et al. | 1997 | Caries Res. 31, 4, 268 - 274 |
| E. coli K88 | Schade R et al. | 1997 | ATLA 25, 555 - 586 |
| Rotavirus | Kuroki M et al. | 1993 | Vet. Microbiol. 37, 135 - 146 |
| E.coli K 99 | Ikemori Y et al. | 1992 | Am. J. Res. 53, 2005 - 2008 |
| E.coli | O'Farelly C et al. | 1992 | Infect. Immun. 60, 2593 - 2597 |
| E.coli | Yokohama H et al. | 1992 | Infect. Immun. 60, 998 - 1007 |
| E.coli | Jüngling A et al. | 1991 | J. Vet. Med. B 38, 373-381 |
| E. coli | Wiedemann V et al. | 1991 | J. Vet. Med. B 38, 283-291 |
| Glycosyltransferase Streptococcus mutans | Hamada S et al. | 1991 | Infect. Immun. 59, 4161 - 4167 |
| Dental caries formations | Otake Set al. | 1991 | J. Dent. Res. 70, 162 - 166 |
| E. coli | Wiedemann V et al. | 1990 | J. Vet. Med. B 37, 163-172 |
| Rotavirus, human | Ebina T et al. | 1990 | Microbiol. Immunol. 34, 7, 617 - 629 |
| E. coli | Schmidt P et al. | 1989 | J. Vet. Med. B 36, 619-628 |
| E.coli | Shimizu M et al. | 1988 | J. Food Sci. 53, 1360 - 1366 |
| Rotavirus, human | Yolken RH et al. | 1988 | Pediatrics 81, 291 - 295 |
| Rotavirus, murine | Bartz CR et al. | 1980 | J. Infect. Diseases 142, 439 - 441 |

## References

Axelsen, N.H., Kroll, J, Weeke, B. (1973) A manual of quantitative immunoelectrophoresis. Methods and applications. Scand. J. Immunol., Suppl. 1, 1973

Behn, I., Hommel, U., Oertel, M., Hauschildt, S. (1996). Kinetics of IgY formation after immunisation of hens with different protein antigens. ALTEX 13 (Suppl.), 18-21

Hesse und Doflein 1935, Tierbau und Tierleben, Volume 1: Der Tierkörper als selbständiger Organismus, Second Edition, Gustav Fischer, Jena, pp.494-500

Holtzhauer, M. (1995) Biochemische Labormethoden. 2. Auflage, Springer, Berlin, Heidelberg, New York, Barcelona, Budapest, Hong Kong, London, Milan, Paris, Tokyo

Mattes, A., Witte, K., Hohmann, W., Lemmer, B. (1991) PHARMFIT - a nonlinear fitting program for pharmacology. Chronobiol. Int. 8, 460-476

Männer, K. Physiologie der Eibildung. In: Scheunert, A., Trautmann, A. (Eds.) Lehrbuch der Veterinärphysiologie, Verlag Paul Parey, Berlin, pp. 541-557

Mittermeier, P. (1995). Das Lipopeptid $Pam_3Cys$-Ser-$(Lys)_4$ - eine Alternative zu Freundschem komplettem Adjuvans (FCA) bei der Immunisierung von Legehennen zur Gewinnung von Dotterantikörpern, Inaugural-Dissertation, Tierärztliche Fakultät der Ludwig-Maximilians-Universität München

Schade, R. Schniering, A., Hlinak, A. (1992) Polyclonal avian antibodies extracted from egg yolk as an alternative to the production of antibodies in mammals - a review

Strobach, H. (1994). Kochbuch der immunologischen Analytik. Radioimmunassay und verwandte Methoden. Georg Thieme, Stuttgart, New York

## Suppliers

**TMB substrate solution - Seramun-Blau:**
Seramun Diagnostica GmbH,
Fährwinkel 8, D-15754 Dolgenbrodt,
Tel./FAX: +49 33767-80911

**TMB-, OPD-, ABTS-tablets (several suppliers are available)**
SIGMA Chemie GmbH,
Tel.: 0800 - 51 55 000
Fax: 0800 - 64 90 000
E-mail: DEorders@eurnotes.sial.com

# Subject Index

The manufacturer's authorised representative in the EU is Springer Nature Customer Service Centre GmbH, Europaplatz 3, 69115 Heidelberg, Germany. If you have any concerns regarding our products, please contact ProductSafety@springernature.com

Printed and bound by CPI Group (UK) Ltd, Croydon, CR0 4YY
25/09/2026
02222891-0003